Biodynamics Decoded

Glen Atkinson

© Garuda Consultants Ltd - v 27
All rights reserved 2025
32 Tauroa Rd , Havelock North
garuda@xtra.co.nz

Hardcopy ISBN 978-0-473-74965-1
Pdf   ISBN 978-0-473-74966-8

CONTENTS

PREFACE

This edition of Biodynamics Decoded has been through several incarnations. Initially the body of this document appeared in 1979, as a university essay. Due to the wonders of computing there will be some of the original text still part of this document. The original text was revised in the early 90s and then again from 1998 to 2000 . Glenological Chemistry takes over the story from there.

This is a large picture which has provided a sense of innate order in what otherwise can appear to be a very chaotic creation. Over the last 32 years of this works development there has been a lot of practical application of the theories outlined here, with some startling results. Some of these are on our website however many of them have never been recorded. I feel now though that I am talking from a 'knowing experience' of this subject rather than the largely theoretical basis from which I spoke in the original "Biodynamics Decoded". The intervening years from its original presentation has seen the basis of this work change very little. The essential premise, that life is a manifestation of energetic forces and that these forces can be influenced by the Biodynamic Preparations remains the same.

Over the years there are many people who have added bits and pieces to this final edition. In their own ways Wendy Palmer, Jana Lyn - Holly, Taryna Neild and Christy Hartlage have all provided a nurturing basis for which I have been able to work and dream the dreams behind this book. Peter Bacchus has been a tireless co worker and now adventurous experimenter of this world view, while Margaret Mary Farr, Menzo de Boam, Rosemarijn van der Slius, Lise Denise, Caroline Lowry and Deborah Williams have all added their knowledge, support and editing skills to this books final form. Rimu Atkinson has played a significant role in recent years, through facilitating the websites manifestation. I thank you all for the parts you have played.

I also wish to acknowledge the role of the Biodynamic pioneers of Drs Steiner, E & L Kolisko, Hauschka, Vreede, Lievegoed and Messers Pelikan, Sucher, Grohmann and Poppelbaum whose work have laid the foundation upon which this work has relied so heavily. Without their gifts this suggestion, that fuses their efforts all together, would not be possible. I have been a collector of what was here before me. All I had to do was to join up the dots and describe what I saw. I trust you will enjoy the result. The Glenopathy website shows where this book has lead to now.

Only small corrections have been made in this 2025 revision, wanting to keep the pleasant naivety of my 30 year old self. A few 'diamonds' from the future fill in some typesetting spaces, and point the way forwards.

Glen Atkinson - 22 May 2025

SYNOPSIS

This method offers an organisational doorway where the principles of Astrology help interpret Steiner effectively, for the benefit of both. Thank you Deborah Williams for this synopsis.

Introduction -   Page 16

Establishing the basic organisational structures of creation.

A brief introduction to Steiner and how Steiner extended the works of Goethe.

Steiner described the world through what he called two-fold, three-fold, four-fold, seven-fold and twelve fold processes. Astrology is built upon the same patterns. The basic 'formula' is introduced, and is shown to be built firstly upon a structure of astronomical principles and then upon Astrological principles. This structure is further developed (and continues to develop) as an overview of a universal theorem, which in conjunction with Steiner's indications, can be applied very practically to all of life.

Astronomical Pictures -   Page 21

One of the basic assertions of the book is that there is an archetypal order upon which all things arrange themselves. Once this order is identified and understood then life processes can be approached in a harmonious and realistic manner. For this to be so then the order can be seen at all stages of existence.

Several photographs from the Hubble Telescope show the electromagnetic shape of a galaxy. This provides us with the form of a double vortex and the horizontal plane.

Further photographs/pictures show how the horizontal plane develops into a lemniscate (or figure 8) along the horizontal plane.

A summary of these forms identifies:
the vertical and horizontal axis, the double vortex and single vortex form, how matter and energy moves down the vertical axis to the centre and back out along the horizontal plane, how matter organises along the horizontal plane according to the electromagnetic fields of the gyroscope and there's a lemniscate pulse which can be observed on the horizontal plane which also organises itself into this Torus form.
8 Pictures and Diagrams

What's There -

We begin the journey through the Astrological theorem with the practical realities of our environment. We start with the immediate environment first. We look at the earth, then the atmosphere and how it developed through the life processes, beginning with green algae and development of free oxygen in the environment.

Then the solar system and Bode's law and how the planets organise themselves according to the electro-magnetic resonance of the solar system.

Then the galaxy. Looking at the various shapes and order that is present within the galaxy.
5 Diagrams

Spiral of Life -

Look at the galactic form. There is a double vortex with a horizontal plane - look at the various shapes which we see in our solar system and our galaxy and how this brings us to the vortex of how the galaxy, solar system, atmosphere, earth and polarity basically sit on top of one another in the shape of a vortex. 5 Diagrams

Spirals and Vortices -

How we look at this gyroscopic big apple/onion form. How this is the overall basis by which order structures itself.
We look at the various levels of the galaxy based creation and how this is imaged also in the astrological model with the zodiac, planets, elements, modes and polarities.
The astrological model as a universal theorem is introduced for the first time. 3 Diagrams

Universal Theorem, -
the 1-3-fold Layers

"As above So below" is imaged in the double vortex of the macrocosmic manifestations and the microcosmic manifestations as lead in before exploring the parts.
The Levels of the Spiral are explained:
Level 1: unity.
Level 2: the basic process of Polarity we see in mythology, psychology (anima, animus etc).
We outline how in a cell for example we see 'singularity' beginning to move thus starting to create a pulsing lemniscate form which eventually develops into two individual spirals. Also we see this polarity functioning between

sun and earth, between cosmic and terrestrial, force and substance activities taking place.

Level 3: how the two-fold polarity becomes the thesis/synthesis/antithesis process in the cardinal, mutable and fixed of astrology and how these are basic laws.   3 Diagrams

The Elements, the 4-fold Layer
How 3 fold becomes 4 fold.
Seeing the Macro and Micro polarity laws
The macro-polarity between the male elements of fire and air and the female elements of water and earth. The micro-polarity which shows up between the fire and earth elements, and the air and water elements is also identified.   4 Diagrams

The Solar System, the 7-fold Layer
How the 4-fold moves into the 7 fold
When the planets are arranged according to the length of their 'sun' cycles an order is discovered which provides a macro and micro polarity structure. The micro-polarities which show up between the inner and outer planets e.g. between Moon-Saturn, Jupiter-Mercury and Mars-Venus with the Sun acting in a harmonising position here.   2 Diagrams

The Galaxy, the 12-fold Layer
The Zodiac and the 12-fold structure.
A brief discussion on the equal and unequal divisions of the zodiac which are found in astrology. The Internal relationships of the zodiac based on the earlier laws work into the zodiac.

Each previous law provides useful insights on how the zodiac 'works' as a practical theorem for life.

The 2-fold law divides the zodiac in the positive and negative signs. Organised according to the planets -7fold- we have an order forming of the zodiac.

The three-fold and four-fold laws are found to be replicated automatically in this order. Several further basic laws of organization are outlined by this action.

Thus we conclude

"For such order to be expressed in this final stage,

all earlier relationships and conclusions about the theorem must also be correct."   5 Diagrams

Evolution on a Pinhead - Page 55

Gives a brief intro to the agricultural lectures and Steiner's work outlining essential understandings of his picture of evolutionary cycles. A brief intro into the energetic bodies and how life is a manifestation of the macrocosmic bodies of the Galaxy, Solar System, atmosphere and Earth, incarnating through the enfolding of the bodies into manifest form.

3 Diagrams

to how it manifests in the inner environment according to the 'micro' polarity.

This provides a clear outline of the way in which the energetic bodies function and we also start to look at the four-fold pattern which shows up in the many different elements of life from the kingdom of nature, the physical bodies, the energetic bodies, the elements, the ethers, the forces, the fruits, the chemical realm (hydrogen, nitrogen, oxygen, carbon), genetics and so on.

6 Diagrams

Level 5   Page 107

The 7-fold - talking about how the inner polarities of (Sun) Saturn, Moon, Jupiter, Mercury, Mars, Venus show up in plant forms, also in the various patterns we see in metals, the organs, chakras and so-on.

We then look at the compost preparations and how they relate to the planets and the associations we can see with using those preparations essentially. We then look into plant forms and the way the planets work according to what Steiner has talked about in the Agricultural course.

Some of the ways that plants are associated to the planets.

13 Diagrams

The Preparations -   Page 116

This chapter sheds new light on the working of the preparations as controllers of the energetic bodies as shown by the Biodynamic Vortex and Steiner and Lievegoed's indications.

A clear interpretation of how each preparation works on the energetic bodies is given.

This perspective has not been put forward in the Bio-dynamic world before and is the basis of the Bio-dynamic essences produced by Garuda. It has allowed Bio-dynamics to move to a whole new level, where it has never been before.   3 Diagrams

Level 6   Page 126

(12-fold) The zodiacs and how they organise themselves and how the 12-fold process manifests

in the mineral kingdom according to the work of Hauschka,

in the animal kingdom according to the work of Kolisko

and we look at planting by the moon using the 5 different moon rhythms.

And then the human kingdom .

14 Diagrams

Which Constellation? -

This chapter offers a unique approach on how to use the 3-fold and 4-fold levels of the Bio-dynamic vortex to understand the constellations of the Zodiac and how therefore which constellations is used to effect plant growth in a particular manner. This is building on the scientific work of Maria Thun of Germany. This methodology allows for very specific use of individual constellations. (Maria Thun only went so far as talking about earth, fire, water, air constellations and this chapter gives an interpretation of each of the constellations.)

We then look at addressing fungal and pest diseases using this system so it allows people to choose what is the appropriate constellation for them in their agriculture.

A further section can be included which details an further understanding of the constellations and plant growth based on a development of Dr Lievegoed indications - very innovative.

The Question of the Zodiacs -

The tropical and sidereal zodiacs in detail.

The polarity between spirit and matter evident in these zodiacs is identified.

A essay on the evolution of consciousness as imaged in our movement away from the constellations to a Sign based Astrology.

1 Diagrams

Equinoctial and Seasonal Zodiacs -

A further development of the chapter at the end of Astrological Science

We identified are the Equinoctial (Cancer > Leo) and Seasonal (Aries > Pisces) zodiacs as the archetypal and manifest zodiacs.

Then explore the Equinoctial Zodiac as the zodiac of the Precession of the Equinoxes and thus the great ages of evolution

Also how the Moon nodes relate to this

What are the 'Houses' of Astrology in regard to the above.

7 Diagrams

Plant Predators -

An outline is given on how to understand pest and disease control within the context of energetic bodies. This provides indications for how we can approach these problems from a new way, not killing them, but by just changing the environment so that they can no longer exist within that environment. Garuda BD Essences are used to achieve this. Case Studies and Pictures are available to support this approach. Again a completely innovative approach to an age old problem.

The Biodynamic Gyroscope

The Vortex exists within a spherical energetic torus, so how can the Biodynamic Vortex be expanded into the Sphere?

The Gyroscopic Periodic Table

A gyroscopic diagram which outlines the main features of Dr Steiners "Agriculture" on the gyroscopic model. This extends the "Biodynamics Decoded" vortex picture into a gyroscope and takes it on to an interpretation of the Chemical Table of elements.

Plant Growth Achievements

An overview of the Garuda BD trails held since 1980—2025

Bibliography

Charts

A4 diagrams. "Biodynamics Decoded"

Thank you Deborah

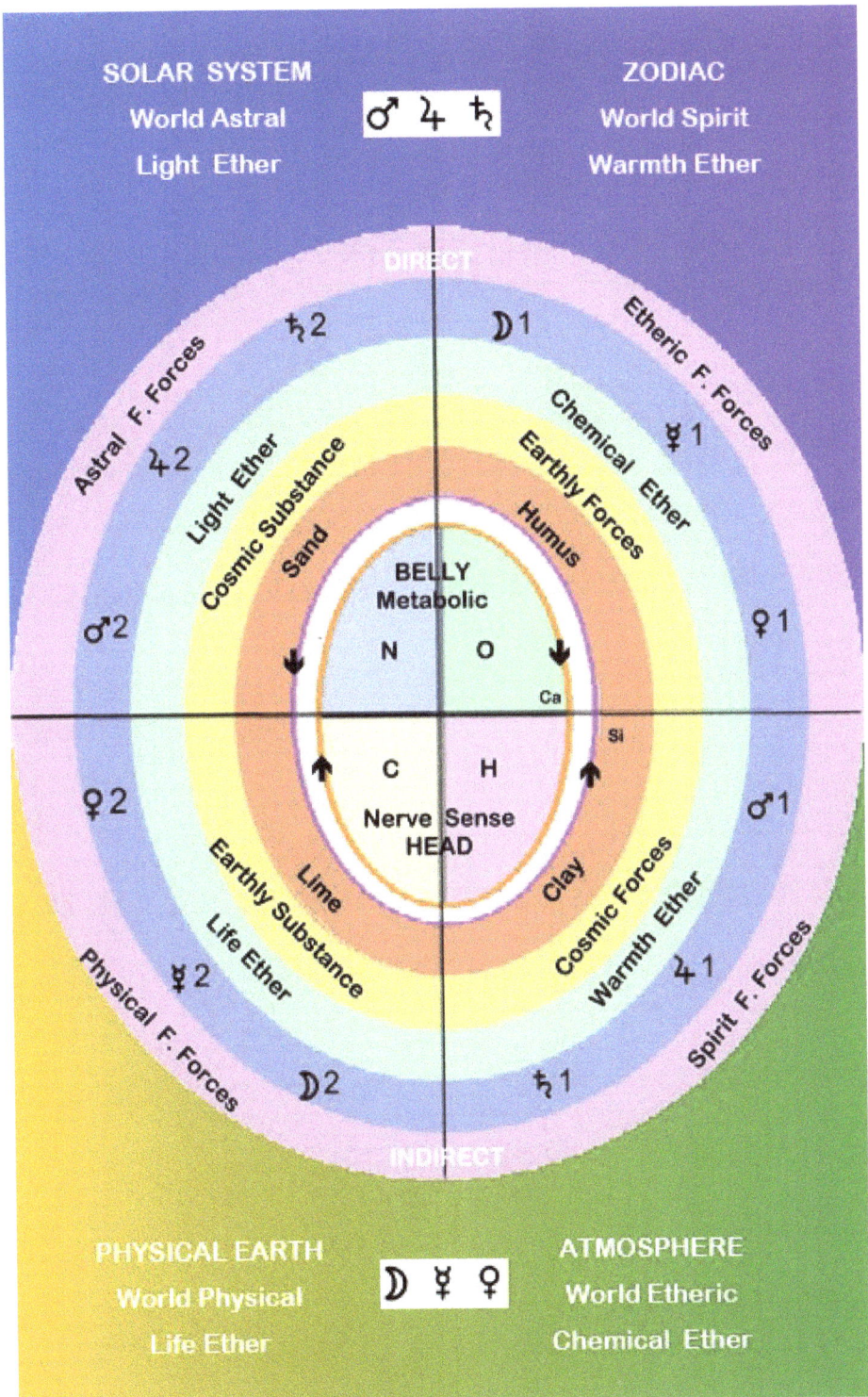

SOLAR SYSTEM — World Astral — Light Ether — ♂ ♃ ♄ — ZODIAC — World Spirit — Warmth Ether

DIRECT

♄2   ☽1
Astral F. Forces   Etheric F. Forces
♃2   ☿1
Light Ether   Chemical Ether
Cosmic Substance   Earthly Forces
Sand   Humus

BELLY
Metabolic
N   O
♂2   Ca   ♀1
Si
C   H
Nerve Sense
HEAD

♀2   ♂1
Earthly Substance   Cosmic Forces
Lime   Clay
Life Ether   Warmth Ether
☿2   ♃1
Physical F. Forces   Spirit F. Forces
☽2   ♄1

INDIRECT

PHYSICAL EARTH — World Physical — Life Ether — ☽ ☿ ♀ — ATMOSPHERE — World Etheric — Chemical Ether

RS's Agriculture Course

INTRODUCTION

Astrology is the science of the interplay between all cosmic phenomena and their corresponding influence on life on Earth.

Anything which comes under this format is Astrology. It is an extremely wide banner which, surprisingly for some, covers 80% of Rudolf Steiners (RS) work.

Both Astrology and Bio-Dynamic Agriculture are large and complex subjects which take serious study to be understood in depth. The aim of this book is to make this study as easy as possible and through their synergy offer a rational explanation for our existence on earth and how to live it consciously.

This book offers a short cut to seeing how the bones of Bio-Dynamic understanding and the order of Astrological understanding form a synthesised basis upon which all the flesh of knowledge is placed. Like a road map this ordering, hopes to help the newcomer chart a quick course to understanding. This journey is worth the effort. Just be aware that this study may well be damaging to your existing perception of reality!

Bio-Dynamics has developed out of the work of Rudolf Steiner, who in turn extended the work of the earlier German poet and scientist W.J. von Goethe. Goethe presented a world view in the mid 1700s which challenged the rationalistic deductionism of Bacon. He asserted that reality can be correctly perceived through the inclusion of subjective experiences of the subject studied, as well as through the objective deductionist techniques favoured by our present scientists. Goethe discovered that by rational observation of one's subject and direct experience through ones imagination and senses, a world of creative metamorphic force can be perceived in plant growth and animal development. It is suggested this structure of force is what is ultimately responsible for the growth and form of the life we see on Earth. The physical forms we see are but the residue laid down along the force's pathways.

Rudolf Steiner and his followers have developed and described this perception in minute detail over the last seventy years, leaving us with a treasure chest of understandings and practical advice to be applied to agriculture and human health. While much of Steiner's information is available in various publications, the student is left with a vast array of concepts which still needs synthesis, before they can become functional in practical use.

Behind what Steiner outlined, an order can be recognised. This order describes a layering of processes which leads to a 'organisation' of all life into a surprisingly coherent pattern. Steiner's layers are the same divisions used in Astrology to describe the polarity, modes, triplicities, planets and Zodiac.

Having established that Bio-Dynamics and Astrology have the same underlying structure allows for the combination of information from both, and in doing so, Biodynamics is organised and synthesised, while Astrology is extended in many practical ways.

There is, however, one further step needing to be made to the layers of both Steiner's Bio-Dynamics and Astrology, and this is to see the layers as a Vortex - the form whereupon the 'independent' layers are organised. The vortex pattern places all the layers one on top of the other and thus provides a basis for clearly establishing the relationships between the layers as well as the relationships between all the parts of each layer.

Using this structural outline as a starting point, rather than an end product, an overview is established early on, which allows new

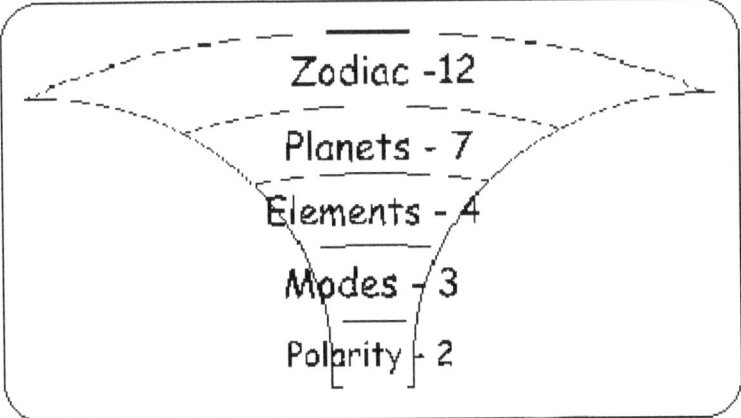

Zodiac -12

Planets - 7

Elements - 4

Modes - 3

Polarity - 2

information to be incorporated and utilised more efficiently, from the outset. This in turn allows for firm conclusions and associations to be easily established. These conclusions can then be further developed using the patterning's and 'laws' present within the basic layered vortex structure.

Working with this Astrological formula and Steiner's indications for the last twenty-odd years has shown me that the Astrological formula is actually the basis of a Universal theorem that provides an in-depth Astrological Scientific approach for working practically with many aspects of manifest life.

Astrology as a system of thought, describes the life processes of our planet and how they are influenced by the greater solar system and galaxy. Its simple formula outlines the structural form and laws active in the creation of all life on Earth . Astrology can do this because it is based on the very organs of our environment, the elements of the earth, all the planets of the solar system and the other stars of the galaxy. This astrological model offers a timeless 'structure of correspondence' useful in cross-referencing seemingly unrelated pieces of information, in a manner that helps to "enlarge the picture". One fact or idea can be associated with other known 'facts' through what can be seen as a universal theorem. Theorems allow for the explanation and comprehension of known phenomena, as well as the projection of new potentials along 'proven' associations. New frontiers can be crossed in the mind before their trial in practical arenas, which forms the ultimate 'proof' of any theorem.

This process of association can be seen as a game. As in any game, we must first learn the rules and understand the bones of the reference system. The second part of the game is putting new information onto this structure; while the third part is the projection of the new information into practical outcomes.

With the Earth in such a state of disrepair, it seems timely this information should be as accessible as possible to as many people as possible. The heart of Steiner's message states that there are 'energetic realities' we must consider in agriculture and Humans, as well as the physical phenomena of soil and climate. If we are again to find health and harmony, understanding these energetic forces and bodies becomes the starting point for any truly sustainable system of agriculture. My experiences have shown that Astrological Science applications may well offer techniques which rival the developments in Genetic engineering for plant control, and evolution.

Rudolf Steiner outlined a perspective to achieve this in his 8 lectures on Agriculture, as well as in other lecture series. He suggested our planet is encompassed by a sea of forces, streaming from the fixed stars (formative forces) and mediated by the planets and our atmosphere. These formative forces go on to be influenced and mediated by the various minerals, crystals and elements of our planet. Thus impacting upon the formation and development of all life forms here. He suggested it is the working of these forces that sustains life.

Bacteria and pests, are only present in any given environment, because of the combination of forces existing in that place at that time. If the balance of these forces is altered, the pests depart for a more satisfactory environment. Biodynamics aims to correctly perceive and balance these forces' activity as naturally as possible, and encourages us to act out of energetic understandings.

A word about my use of the word energetic. I use this word in the most practical of ways. Spirit forces are the forces coming from Stars. Their long duration causes a standing wave throughout creation, which forms species as they enter into manifestation. This external energy enters into all life forms for periods of time. The Spirit's actions can be seen through its influence on external manifestations. It is one of the Energetic Bodies. The term 'energetic bodies' I use to mean all the three bodies, not seen directly with the average human eye. The spirit is not used to mean anything one believes in, or has faith in. As Star Forces it acts into life very practically. In keeping with Steiner's usage, I use the word "Ego" to indicate that part of humans which comes into action when the spirit incarnates into them. The word "ego" is used rarely, but is used to mean the lower self, as in "egotist".

This work does not attempt to be a complete book of Bio-Dynamics. There are many titles that cover practical applications and any reference information alluded to in this work is detailed in the bibliography for those wishing to further study titles that expand topics only touched on here. Nor does this book purport to be a complete guide to Astrology;

The intention of this work is to point the way of this new development: many of the pictures, relationships and diagrams presented here are meant to act as a starting point for your own contemplation. They are doorways to understanding. This work does not make any pretensions to being ' scientific'; in the normal sense, if anything it is an artistic attempt at bringing the 'Wisdom of the Ages' into a modern context for the benefit of Agriculture, Humans and the health of the planet Earth. The practical work and innovations already developed from this thesis and presented in 'Case Studies' have proven the theorems worth and validity. Their reality must be addressed and otherwise explained by any sceptics.

This information is offered as a sharing of the picture and experience of order and harmony that Biodynamics and Astrology have brought to my life on this planet 'Earth'. I hope you are broadened, challenged and reassured by it in the same manner that it has influenced me This is a continuous journey of exploration with new developments showing themselves whenever I have the time to pursue them. This book is only the beginning of an ongoing project, which will be added to over time by myself and hopefully others.

ASTRONOMICAL PICTURES
Establishing Archetypal Patterns

One of the basic assertions of my Biodynamic world view is that **there is an archetypal order upon which all of life arranges itself.** Once this order is identified and understood then life's processes can be approached in a more harmonious and realistic manner and the order can be seen at all stages of existence. To find the archetypal forms of life we begin our study with the largest organisational forms of our existence - the stars and galaxies within which we live. The following pictures are from the Hubble telescope operated by NASA. They provide us with Astronomical pictures of how matter and then lifeforms come into formation.

This blue picture is of a star in the constellation of Virgo. The defining feature of this star is that we can see a clear vertical axis and the formation of a horizontal disk. This is a young star and still in a predominately gaseous form, thus we can still see the vertical column as gases are sucked in the vertical axis and squirted out along the horizontal plane. What we are not seeing here is the electromagnetic fields which are active in the formation of what we are seeing. The basic structure of these fields is detailed in diagram 2a.

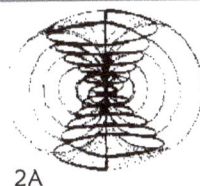

2A

A common feature of galaxies, stars, planets and all other things is the development of an electromagnetic field based upon this pattern. This electro-magnetic field is generated by the 2B immense speeds these stars spin at (our Sun spins at approx. 64,000 mph). As any substance spins, it develops two vortices which begin to suck more substance into a central mass. In

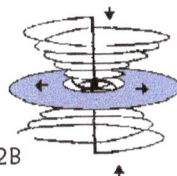

the case of stars this substance is hydrogen. Once the hydrogen reaches a certain point of density and pressure it ignites and begins to express, or explode, matter outwards along its horizontal plane (See diagram 2b).

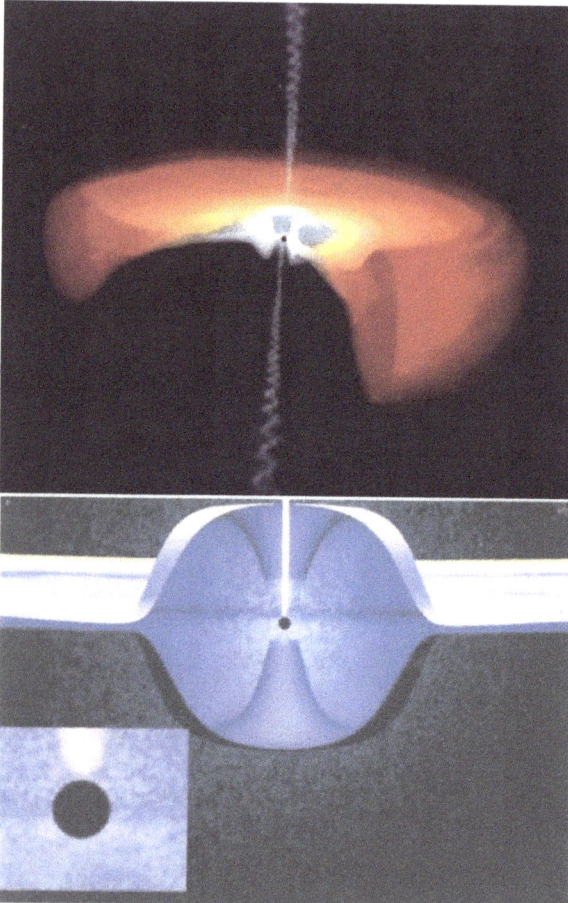

As the star cools it enters into the next p h a s e   o f manifestation - the vertical axis becomes less visible, while the h o r i z o n t a l   a x i s b e c o m e s   m o r e defined.

The picture on the top left is of a cooler star than the blue one above and has a cut away section so we can see the details of this   stage.   The features to note are the double vortex in the middle the small width of the first two t h i r d s   o f   t h e horizontal plane, and the bulbous quality of the outer section of the horizontal plane.

The bottom picture is the centre of the previous picture. Here we see clearly the vertical double vortex at the centre of all gyroscopic forms. The black dot in the centre is the supposed black hole at the centre of all gyroscopic stars and galaxies.

In the next stage of development in a galaxy formation, the vertical axis becomes invisible while the horizontal plane differentiates according to the nature of the electro-magnetic fields of that gyroscope. In the case of galaxies, spiral arms develop; in the case of

Stars, planetary systems are formed, and around the planets Moons form. The harmony of this division is described by Bodes Law, which is shown on page 28

In the early development of the horizontal plane another phenomena can be observed.

When looking at the bottom picture, of a supernova , the expansion of the horizontal plane appears to reach a stage where a certain boundary is reached - possibly the edge of its electro-magnetic field - before it turns back in on itself; thus creating an internal pulse based upon the form of a lemniscate, or "figure-eight".

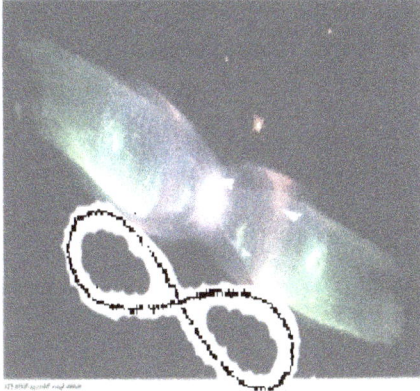

Summary

The forms we can see in these pictures are:

1) Vertical and Horizontal axis.

2) Double vortex & single vortex forms.

3) Matter and energy moves down the vertical axis to the centre and back out along the horizontal axis.

4) Matter organises along the horizontal axis according to the electro-magnetic fields of the gyroscope.

5) A lemniscate pulse can be observed on the horizontal plane.

These forms, are the basis of the immense gyroscopic beings in which we live, are also the basic organisational patterns of the life forms

WHAT'S THERE

"... when we want to understand the plant, we must bring into question not only plant, animal and human life, but the whole Universe. For life comes from the whole Universe not only the Earth. Nature is a unity and her forces are at work from all sides. He who can keep his mind open to the manifest workings of these forces will understand her.

" Rudolf Steiner Pg 70, 1938 Agriculture. (1)

Astrology is ultimately a theorem developed from the reality we live in. It is therefore best to begin our journey through the theorem with the practical realities of our environment, discover the archetypal patterning upon which this stands and then look again at manifest life.

Rudolf Steiner's picture, which is not unique, presents life functioning on various levels of activity all at once. Somewhat like a multi-levelled game of chess. He continually referred to the two-fold, three-fold, four-fold, seven-fold and twelve-fold organization of life processes. Astrology uses these same number divisions to organize information. Physically we can explore similar layers through investigating our immediate big neighbourhood, which is bigger than most people might initially think...

The Environment

When we grow crops, we need to look at our environment. The first step is to look at the immediate physical quality of our land.

The Earth. -The Physical realm

The quality of the soil meets us first, the soil drainage, nutrient status, its structure, the type of soil, sandy or clay and the water-holding capacity. Is there enough water? Do we need to irrigate? Does it drain? Do we need to mulch or balance the minerals? These are all aspects of the physical environment we consider. We then consider the way light and warmth work into our soil, and the garden as a whole. What is the rainfall and when does it come? These questions form the climatic level of our considerations.

We look to see how our trees are planted in the environment. Does the neighbour have a big gum tree sitting in front of the garden? Are you in an urban situation or by the sea? Then we look at the wind. How strong is it, how often does it blow? Or does the wind blow unblocked from one direction? From here we look at the larger geographical picture. How are the hills arranged? How does this affect the rain, wind, light etc. ?

We see this as a bigger picture and look further still to see that in New Zealand, for instance, we are living on an island on the edge of the Pacific Ocean. This in itself brings a certain climatic situation. We are aware of being a long island with masses of water, and a central mountain range, all of which works mostly to our advantage.

The next step is to see we are part of a Planet. A large ball of mass which has a unique area around it, the atmosphere.

The Atmosphere -The realm of the Elements.

Our atmosphere is a wonder which should not be over-looked. Its wonder lies in the fact that it was created by the Earth itself over our 5 billion year history. It is from the life processes of the Earth that oxygen was initially produced and released into an environment, which would otherwise be toxic to us. The earliest form of life was the Blue-Green Algae living just below the surface of the oceans. These were the first life forms to develop and produce oxygen. Their existence manifested from the interplay of the four elements of warmth, light, water and the salts and minerals of the Earth found only in our atmosphere.

The cosmic rays of light and heat met the earthly substance of water and salt minerals in the oceans. At the critical interface of the ocean surface the Blue-Green Algae developed. From their production of oxygen over millions of years the Earth's surface eventually oxidised. Thus oxygen ceased being locked up to the same degree. Our atmosphere became rich enough in oxygen for other life forms to begin their development.

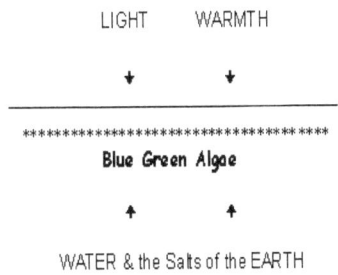

```
        LIGHT      WARMTH


          ↓          ↓
     _____
     *********************************** ****
           Blue Green Algae

          ↑          ↑

     WATER & the Salts of the EARTH
```

As evolution continued the animals in the oceans developed. Through their shells they concentrated and released calcium carbonate and formed the first animal proteins through combining nitrogen with carbohydrates. The great majority of calcium on earth, 28% of the Earth's surface, has built up from the deposits of the sea animals or deposits from other life processes of the Earth itself. Calcium also deposits directly onto the Earth as cosmic dust from the Sun. We will see later the importance of calcium and oxygen as the anchor on Earth of the life-giving forces.

For over 400 million years Oxygen has built up our atmosphere, which now extends up to the ionosphere. Between the ionosphere and the stratosphere we have the ozone layer. This became an important membrane around the planet, reflecting the harmful gamma and ultra-violet radiations from the planetary and fixed star regions. These forces damage life processes.

Within this sphere is the realm of the elements where light, warmth and moisture dwell. Out in space it is dark and as soon as the sun's radiation hits the ionosphere light and heat are generated. These elements all live within this sphere.

Looking at our Earth from the outside we can see it spins on its axis. It moves at 1000 miles per hour at the equator. It moves along a predictable path around an even bigger mass, the Sun, at a speed of 66,000mph. So we come to see the Earth is only a part of a larger environment - the Solar system.

The Solar System - The realm of the Planets.

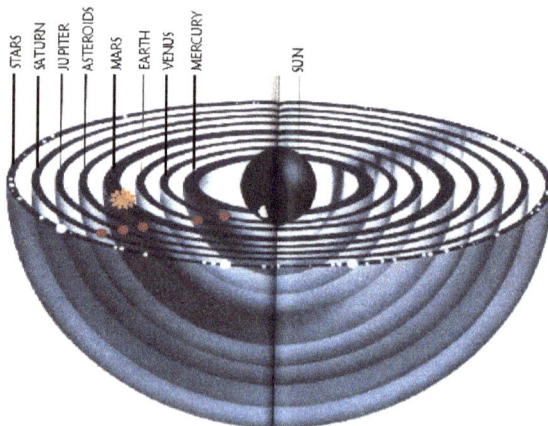

The Solar System Onion

Astronomy tells us the Earth is more than a ball of rubble moving through space. It lives in a precise relationship to all the other planets in our environment. We have the Sun in the middle with a very concentric expansion of planetary spheres taking place. The actual planets all manifest along a horizontal plane. Nearly all are within a five degree arc of the sky. Pluto is the only one which has a large arc which is around sixteen degrees.

This is a very ordered Solar system. To quote Einstein " God creates according to mathematical law". We have the various planets lined out in an order and at distances which can be rhythmically calculated by Bodes Law. (pg. 28   ) This is the formula, rather than a law, for the placement of the planets within the solar system.

From this formula we can predict where the planets will be placed from the Sun according to Astronomical units. Uranus, Neptune and Pluto were discovered using this 'Law' even though the last two are not in their exact positions. This lack of accuracy in their position still leads to the speculation there exists yet another planet Trans Pluto (Persephone) beyond Pluto.

Neptune and Pluto's path could be seen as being interrelated, as one cycle. In our present time Pluto is closer to the Sun than Neptune. Their 500 yearly conjunctions coincide with every second time Pluto reaches its aphelion (furthest point from the Sun). Their cycles and effects on human psychology and history also suggest a strong inter-relationship.

According to Astronomical observations the solar system is essentially a great big spinning top bathed in an electro-magnetic field. As any one of these planets moves around within this sea of electro-magnetic force, so they push and pull all the others within this area. As the planets move, so do the electro-magnetic fields which hold matter in place. Matter moves as if it were sand in the bottom of a bottle. The planets are the only moving part of our Astronomical reality, in the small context of the length of our lives, so this becomes a very significant level of existence. Their realm is very ordered, inter-related and predictable.

Pluto takes 248 years to complete one orbit around our Sun. Neptune takes 165 years, Uranus takes 84 years, Saturn takes 29.5 years and

Jupiter takes twelve years. Within this number series it can be seen that Jupiter revolves seven times for one Uranus cycle and fourteen times for one Neptune cycle. Saturn takes approximately 3 cycles for one Uranus cycle and 6 cycles for one Neptune cycle and so on.

As the Solar System is a giant, electro-magnetic spinning field it can be seen that as the planets move in an organised harmony, any movement causes a rhythmic ripple- effect which impacts upon the rest of the structure of the Solar System.

Diagram 2—Bodes Law

Planet	Theory	Actual AU*	Cycle
Mercury	4-10=0.4	0.4	87.97 days
Venus	3+4-10=0.7	0.7	224.7 days
Earth	6+4-10 =1	1	1 yr
Mars	12+4-10= 1.6	1.5	1.88 yrs
Asteroids	24+4-10= 2.8	2.8	4.6 yrs
Jupiter	48+4-10= 5.2	5.2	11.86 yrs
Saturn	96+4-10= 10	9.5	29.46 yrs
Uranus	192+4-10 = 19.6	19.2	84.02 yrs
Neptune	384+4-10 = 38.8	30	164.78 yrs
Pluto		39.5	248.4 yrs

*AU =	Astronomical Units

The Galaxy - The realm of the Fixed Stars

The next step on our journey through our environment takes us past Pluto and out into the vast distances of our Galaxy. Our Solar system, in its contained spinning motion, hurtles through space in the direction of the constellation of Hercules, at the speed of 48,000mph.

We are moving around a galactic centre in a motion that takes 240 million years to go once around. The Galaxy has a radius of 50,000 light years. Hence here on Earth we are moving at a combined approximate velocity of 115,000 miles per hour (Earth's spin - 1000 m.p.h., Earth around the Sun - 66,000 m.p.h., Solar system around the Galaxy - 48,000 m.p.h. = 115,000 m.p.h.) At this speed one must speculate as to whether we are actually being pushed onto the planet rather than pulled, as current gravitational theory suggests?

It is interesting to note that with our eyes we can only see the planets out to Saturn. When it comes to the Galaxy, the stars we see with our eyes are mostly contained within our Galaxy. While they are moving slowly apart, they appear (in our life time) to remain in the same relationship to each other and our solar system. The Stars we see are either Suns or other systems filled with Suns generating immense amounts of radiations, constantly. These Stars form the basis of the continuous stream of cosmic forces within which we live.

The galaxy is an interesting form. It is in a spiral shape with a centre

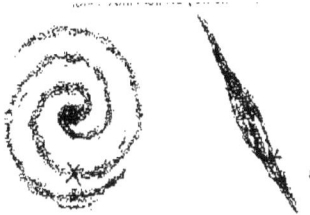

and two arms spiralling out. Our solar system is on one of the outer arms about three-quarters of the way out from the centre.

Our Milky Way galaxy appears to be a disc shape, and is most often represented this way. What we can imagine is a flat plane spinning in space. However this is not the whole story. The Galaxy is actually a sphere shape rather than just a flat plane. We are best to start by visualizing an egg/apple shape. The flat plane we see is just where the

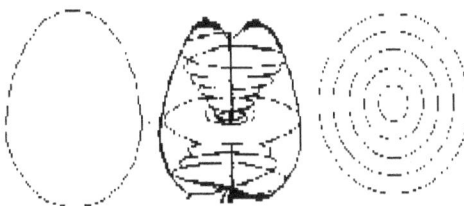

matter coagulates. If we cut an apple in half through the middle (not top to bottom) then the Galaxy disk is on that horizontal plane. The rest of the 'Apple' is made up of the electromagnetic force fields holding all the matter together. To go one step further it would help to imagine the inside of the Apple as an onion, with its multi layers or spheres.

Now, on the horizontal plane there are areas of matter and gaps. Sit with this awhile, as it provides a start for gaining a truer picture of what our galaxy actually looks like. The constellation of Sagittarius marks the direction of the centre of the galaxy in the sky. If you look up in the sky towards Sagittarius you will see the bulge of stars.

Our solar system spins on a different angle from the galactic plane - about 85° different. In doing so we create another self contained spherical dimension, within the body of the galaxy, which has the same egg/apple/onion form as the smaller structure. All we see of the solar system are the planets lined out on a flat plane, just as when we look at the galaxy. We only see the horizontal plane, not the energetic fields.

Beyond our galaxy we form into a local group of eight galaxies, which is said to revolve around the star Alcyone in the Pleiades. This group in turn is part of a super-cluster of such grouped galaxies. These in turn are part of larger groupings called "globular bundles". These represent very big distances in space. Our local super cluster is approximately 100 million light years in diameter.

The galaxies all appear to be moving away from each other. Therefore our local spiralling ordered form, is our Galaxy - 100,000 light years in diameter. I consider this the edge of our immediate organized environment. Hinduism's creation story has Vishnu lying in and Ocean of milk, as their description of this. Four headed Lord Brahma arises from a lotus flower, growing from his navel.

THE SPIRAL OF LIFE

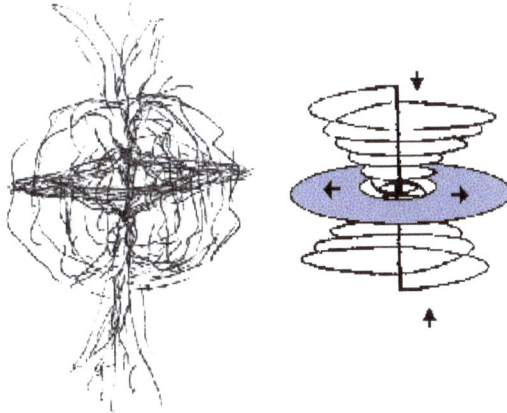

The flat spiral form of our galaxy (see pg. 23) is the end product of a long process based upon motion.   If everything in creation was stagnant then the galaxy would not have formed.

However creation is not stagnant. We assume an event called the Big Bang took place some fifteen billion years ago and that since that time, matter and gases have been propelled outwards from a combusting galactic centre.

This basic gyroscopic form has a set of creative processes. From this we can observe several individual forms which make up the archetypal structure.

There is the Cosmic Egg,

The Apple, with its vertical vortices and accompanying electro-magnetic fields, as well as the Flat Plane which is organised in both circles and spirals.

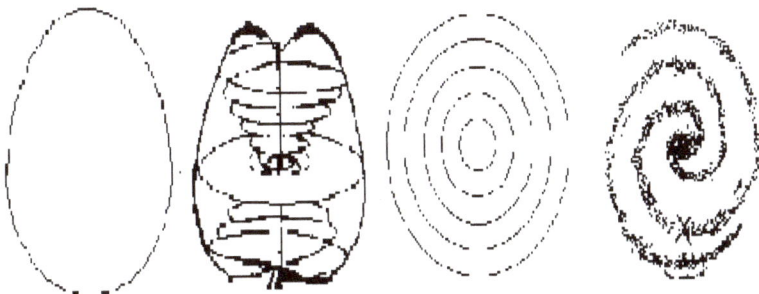

All of these forms, in turn, have inner forms which can be used as archetypal structures.

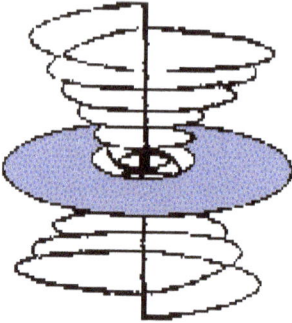

Here we can identify a double vortex form (on the vertical plane), which in turn can be reduced to a single vortex form.

Due to the spiralling and circling effects we have identified in the flat plane created by the electro-magnetic fields, we can deduce there to be layers manifesting within the vortices. So we can add the layered vortex to our basic forms.

In the previous chapter, we found that our reality is made up of living beings upon a planet, which is circled by an atmosphere, which exists within a solar system which in turn exists within an organised galaxy. This then provides us with a picture of the layers of the vortex stretching from the Earth's surface to the edge of the galaxy.

This image provides a picture of the Cosmic Apple from the outside top perspective (right) and the sideways view (left).

On the right we look down the vortex and onto the horizontal plane. This provides us with a picture of the activities related to the different spherical 'onion' ring harmonics as they move to the centre.

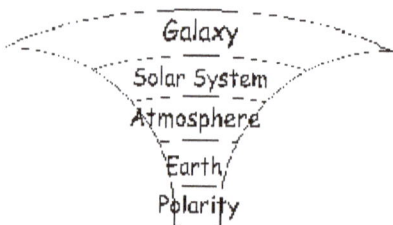

Galaxy
Solar System
Atmosphere
Earth
Polarity

Galaxy = Signs
Solar System = Planets
Atmosphere = Elements
Earth = Modes
Polarity = Sexes

This process of identifying archetypal forms, provides us with several structures we can apply to various aspects of life to identify order in a seemingly endless chaos.

Within the gyroscope there are two spiral forms of note. The spiral of the horizontal plane which manifests as an exponential spiral and the vertical spirals which show up as an even spiral. These two spiral shapes show themselves in many forms of life on Earth, most notably in the two spiralling forms of sea shells. Spiralling forms are found everywhere in nature. The leaves spiral up the stems of the plants and the movement of water is through spirals, while our DNA structure also spirals.

The gyroscope and its secondary form the spiral can act as special 'form keys' for understanding how life organises itself. It seems appropriate that if these are the primary sustaining structures of life from the smallest algae and DNA to the largest galaxy then there has to be relevance in exploring the physics of these forms and to use them as archetypal structures for arranging our understandings.

For the purpose of examining manifest life, it is the exponential spiral of the horizontal plane of manifestation which holds special significance. Spirals go through several levels. We saw earlier how our creation can be seen to move through a spiralling process. Also that each layer of this spiral is actually an independent gyroscopic being in itself, expressed as the galaxy, solar system and Earth. This independent gyroscopic motion at different polar angles to each other, suggests each gyroscope represents independent dimensional realities. Thus moving from one layer of our archetypal vortex to the next represents a shift from one dimensional level of activity to the next. As we are dealing with a wholistic creation there is ultimately no complete separation of any layers. What we have is one large and complex activity manifesting on different dimensional levels

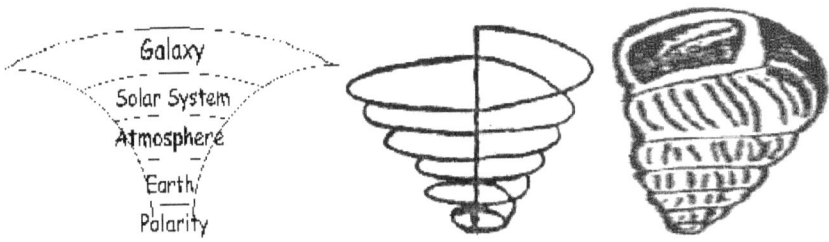

Galaxy
Solar System
Atmosphere
Earth
Polarity

simultaneously. While there is constant interplay between levels in every direction, for simplicity's sake it can be said that as we move up the vortex an activity present on one lower level will manifest at the higher levels, however in a more complex manner.

While the spiral expands through centrifugal forces, it also draws into itself through centripetal forces. The speed and height of the spiral are a manifest picture of the balancing of these two forces. Here we see a harmony between expansion and contraction in continual motion. Life is a dynamic moving interplay of force. As the nature of the force changes so the material matter will follow.

Looking at our environment we are presented with a picture of four primary spheres or levels of activity, outside of our selves. These are the Earth itself, the atmosphere, the solar system and the galaxy. These levels of creation can be seen as the macrocosmic picture of our environment and the first four levels of our spiral of life. The fifth comes when we see that higher life on Earth, more often than not, is divided into male and female sexes. For physical creation to spark in the higher forms of life on Earth, one male and one female being still need to reproduce sexually.

Each of these levels of activity - the fixed stars, the planets, the atmosphere, the physical earth and the polarity of the sexes therefore become the layers of an 'Archetypal Spiral'.

While it is OK to see the layers as being one on top of the other, it is more appropriate to visualise the different levels as being a development of the layer before them in the same spiralling form. Observe sea shells, they have continuous lines around the spiral as well as vertical lines across the spiral.

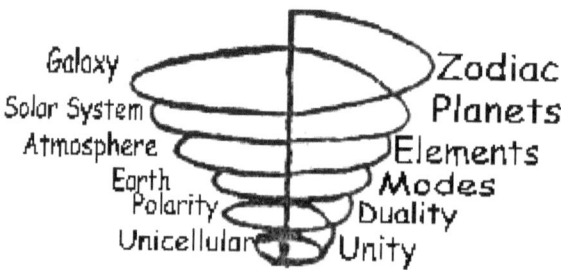

Galaxy — Zodiac
Solar System — Planets
Atmosphere — Elements
Earth — Modes
Polarity — Duality
Unicellular — Unity

Environment   **Astrology**

Astrology uses the same basic elements for describing reality as we find in our environmental spiral. For thousands of years it has described these divisions using the names Polarity, Modes, Elements, Planets, and Constellations.

Each layer of the spiral has a different number of parts. The Astrological world view outlines the inner workings of each layer thoroughly. Level 2, the sexes, has 2 parts; Level 3, the Modes, has three parts called Cardinal, Fixed and Mutable; Level 4, the Elements, has 4 parts - Fire, Air, Water and Earth; while level 5, the Planets, has seven parts - Saturn, Jupiter, Mars, Sun, Venus, Mercury, and the Moon; Level 6 has twelve parts commonly known as the twelve signs of the Zodiac.

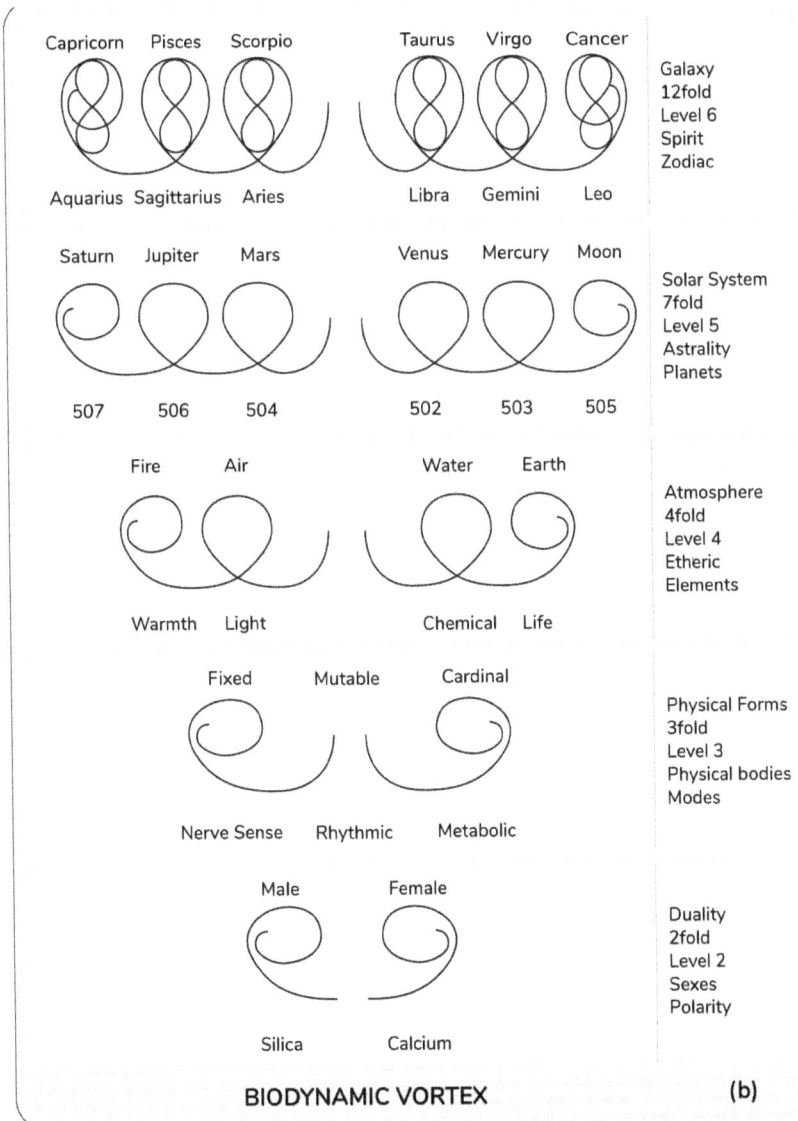

Capricorn	Pisces	Scorpio	Taurus	Virgo	Cancer	Galaxy 12fold Level 6 Spirit Zodiac
Aquarius	Sagittarius	Aries	Libra	Gemini	Leo	
Saturn	Jupiter	Mars	Venus	Mercury	Moon	Solar System 7fold Level 5 Astrality Planets
507	506	504	502	503	505	
	Fire	Air	Water	Earth		Atmosphere 4fold Level 4 Etheric Elements
	Warmth	Light	Chemical	Life		
	Fixed	Mutable	Cardinal			Physical Forms 3fold Level 3 Physical bodies Modes
	Nerve Sense	Rhythmic	Metabolic			
	Male	Female				Duality 2fold Level 2 Sexes Polarity
	Silica	Calcium				

BIODYNAMIC VORTEX   (b)

35

THE UNIVERSAL THEOREUM

Astrology at first glance can seem to be an unstructured mass of symbols and concepts: the sign of Aries, the planets Saturn, Jupiter etc. Are you an earth sign, or are you a cardinal sign? These are all pieces of information that can be confusing to bring together. Astrology has a definite structure: The basic pieces of this layered vortex are named the polarities, the modes, the elements, the planets, and the zodiac. Each layer represents a plane of activity, which is whole in itself with its own inner relationships, while being part of the greater whole.

The next step in our journey is to explore each layer to establish their inner qualities and laws. As Rudolf Steiner's lectures and Astrology both show, these laws act as an archetypal order. We can learn the nature of the parts by rote without any sense of order, or we can learn the rules and order that stands behind matter and then apply them to whatever form of manifestation we are confronted with.

At this stage we are seeking for principles. In later chapters we will see how these principles are related to Cosmic and Manifest forms. The diagram below shows how the double spiral becomes a picture of the Cosmic and Earthly manifestations of the principles found within the archetypal vortex. This is truly "As Above So Below".

With all the parts, being part of a moving form, one stage leads into the next. This is a continuous state of development from simple to

World Ego — Zodiac
World Astral — Planets
World Etheric — Atmosphere
World Physical — Earth
Polarity — Sexes
Oneness — Atom
Sexes
Physical Body — Mineral Kingdom
Etheric Body — Plant Kingdom
Astral Body — Animal Kingdom
Ego - Spirit — Human Kingdom

more complex. One layer develops naturally into the next. Vortices by their nature have a twirling exterior and a calm passive centre space. In the whirlpool of water this centre space is where the dross of matter accumulates, in the tornado or cyclone it is an area of calm. In the Human this calm space is found when our Spirit is properly incarnated, and we become 'centred'. These examples suggest the centre of the vortex is the place of universal creativity where manifest reality exists in peace, amid the turmoil of the forces which hold it in place. In fact, the quality of the middle is an expression of the dynamic interplay of all the vortex processes, so as they alter their relationship to one another so the centre changes.

LEVEL ONE - Unity

The first level of the spiral is seen as a state of oneness. This represents all states of oneness from the universal oneness to the unicellular algae to the atom. Either way it is where all 'life' begins and is undifferentiated.

LEVEL TWO - The Polarities

At level two polarisation occurs. Polarisation appears to be the basic criteria for creative manifestation to take place and be maintained. We see this first in the polarisation into positive and negative poles within the electro-magnetic development of even a drop of water moving through the atmosphere. All chemical elements are electrically charged either positively or negatively which acts as the bonding basis for more complex molecules. D.N.A. , the basic molecule of life on this planet, carries this polar image within its two strand structure. The well-known divisions of light and dark, heaven and earth, male and female are images of this process. In mythology this level gives rise to most of the primary creation myths of the earth . In psychology, the introvert, extrovert types and Jung's Anima and Animus divisions follow this level. This level, carries the image of the single cell beginning to pulsate, setting up a movement that becomes an active pulsing lemniscate. This increases until two whirling vortices of energy are established. One spirals outward and the other inward meeting in the middle. The interplay and subsequent creative tension gives rise to the spark of creation.

In Astrology these poles are named after the predominant planets in our sky. The assertive male pole is associated with the Sun, while the receptive female pole is associated with the Moon. Have you noticed

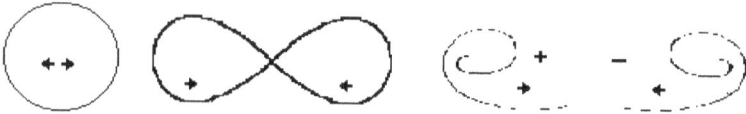

that while the Sun and Moon are actually different sizes, when viewed from the Earth they are the same size? This primary polarity can also be imaged as a balance between Sun & Earth, cosmic & terrestrial, force and substance. Every situation we observe can be seen as an interaction of these two primary poles.

This is the basic law of manifestation and it extends into all the other levels of the spiral. The deeper into matter we explore, the stronger the need for polarisation.

LEVEL THREE - The Modes

The third level of the spiral is known as the modes. At this stage the polar opposites meet and develop a middle ground. The theory of Thesis, Antithesis and Synthesis describes it very well.

In Astrology the male pole moves anti clockwise towards the centre while the female pole moves clockwise towards the same centre. The middle is a direct manifestation of their interplay. The middle is said to be both expansive and contractive, moving with the need of the time.

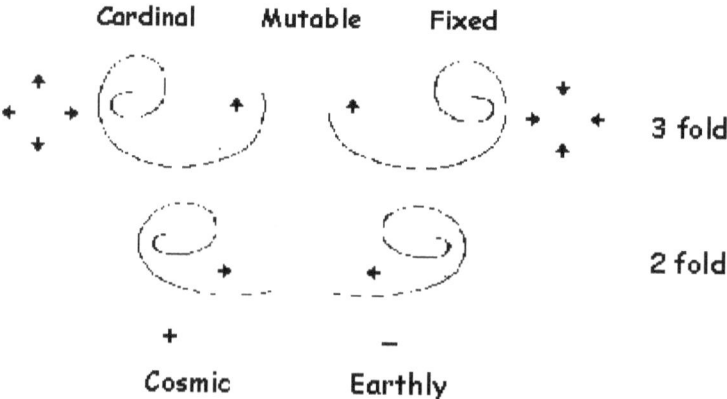

Astrologically these are described as Cardinal (Thesis) the impulse towards assertive action, Fixed (Antithesis) the impulse to remain with what already exists, and Mutable (Synthesis) The impulse to acquiesce to find the middle ground of rhythmic harmony. In the interlevel cross-reference the assertive, cardinal pole has similarities to the male Sun pole of the earlier level, while the contractive, fixed mode is associated with the female Moon pole. The mutable mode is the harmony created from merging of both.

This principle is illustrated in the division of the work place. The directors of a company are the Cardinal role. They determine the direction, look at new opportunities and need to innovate to reach the goals they have set. The workforce are the Fixed pole of the business. They have fixed routines, with predetermined breaks with fixed holidays and remuneration. The Mutable pole is the management team in the middle. Mutable people are the gap fillers -when something needs to change or a gap in the order of life opens up then the Mutable person easily turns their hand to the task, for a short period of time. Their life blood is variety. Cardinal's is new horizons, while the Fixer's is routine.

Rudolf Steiner allocated the opposite correspondences. To his thinking, Fixed is associated with the Cosmic nerve sense pole, shown in the contracted focus of thinking, while the Cardinal processes are associated to the volatile limb metabolic Earthly processes. Both associations are right. One application (Astrology) is talking of Human psychology, while RS is talking of organic growth processes.

The main thing to see here is the duality and difference of a primary Archetypal processes (RS) and a process at the end of Manifestation — Human psychology. See pg 86.

Archetypal Order   **Human Psychology**

LEVEL FOUR - The Elements - 4 fold

Level four of the spiral vortex finds its external basis in the atmosphere of the earth. Astrology uses the four elements found in the atmosphere, Fire, Air, Water and Earth to describe the activities on this level. These are the very substance upon which life began and from which most life forms are made.

The creation of this level occurs due to interaction of the polarity of male and female (Level 2) giving rise to a third component which is a synthesis of their extremes (Level 3) which gives rise to the Modes. As this burst of energy continues, the Mutable phase divides into two sections which then provides us with four pieces. These are the elements of Fire, Air, Water and Earth.

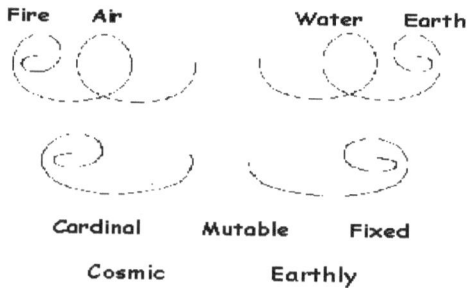

Each element has a quality that makes it unique, yet they follow on in their development from the level below them on the spiral. The 'outside' elements of Fire and Earth hold the extreme qualities of those before them, Cardinal and Fixed respectively. The middle mode, Mutable, divides itself into Air and Water. This can be described as a development of the positive and negative indicators as follows.

Fire ++   Air +-   -+ Water -- Earth

Fire is assertive ,   impulsive
and expansive in all directions at one time, with its activity generating warmth. The choleric temperament is a mark of its action and Jung called it the Intuitive type because of its tendency to act spontaneously without much rational consideration of a situation.

Air is also expansive, however more single pointed in quality. It uses rational thought as a medium of action, rather than brute force and is fine and light in nature. The element of air, is oriented to more social interaction and communication, rather than assertive action. The Sanguine butterfly type temperament is associated with Air. Jung classed this as the Intellectual type as the mind often dominates all other functions.

Water is fluid , passive and emotional in comparison, exhibiting qualities of compassion and nurturing. It will assume the form it is contained in and always moves to the lowest point of motionlessness. The phlegmatic temperament carries these qualities. Jung called this the emotional type due to these considerations dominating other facilities.

Earth is solid and supportive tending towards being cold and attracted to earthly sensuousness and security. Practical and stable in nature, it is associated with the melancholic personality. The Sensation type in Jungian terms is due to the love of physical touch and sensual pleasure.

Elemental Association

Individual elements associate in harmony and discord with each other. The activity and relationship of the elements is an extension of earlier stages, yet there are now two different relationships to consider. The Macro polarity that is found most often in the External World, and the Inner polarity found within Living processes.

Macro Polarity

The 'macro' polarity is a continuation from level two - Cosmic and Earthly, which shows here as the brother/sister relationship between the expansive elements of Fire and Air balancing a similar relationship between the contractive elements of Water and Earth.

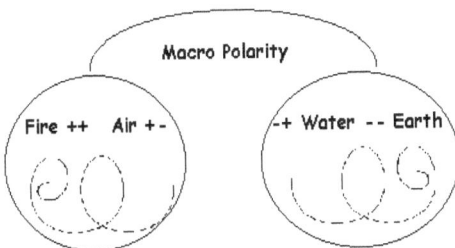

Macro Polarity

Fire ++   Air + -   -+ Water -- Earth

Fire and Air generally work well and support each other, while Water and Earth do the same.

Inner Polarity

The second polarity structure arises as a result of the increasingly complex developmental process of the Astrological vortex. These polarities arise as a result of the **internal polarisation between the opposites of each pole.** Fire ++ and Earth -- form a creative polarity, maintaining and mimicking the extreme qualities of the Level Two, Cosmic and Earthly relationship. The assertive and sometimes impulsive urges of the Fire element are balanced by the practical common sense perspective of the Earth element.

From the mutable mode of level three, we have the development of Air and Water. These elements function supportively as well. The intellectual, rational tendencies of Air can be seen as balancing the more emotional and sometimes irrational quality of Water.

It is at level four that the Gyroscope comes into formation and is therefore a useful form for appreciating the relationship of these elements. It is within the Atmosphere that life becomes sustainable. It is also when the gyroscope become fully formed and its motion balanced that it becomes a perpetual motion device. Our Earth, Sun and Galaxy have been running for some billions of years afterall. It is at this stage that the two primary poles of the vertical structure interact to create the horizontal plane. The internal pulsing within this middle sphere becomes such that two separate spheres of activity become differentiated within it. The gyroscope is formed. It is apparent that the vertical pole is sustained through the tension of Fire and Earth, while

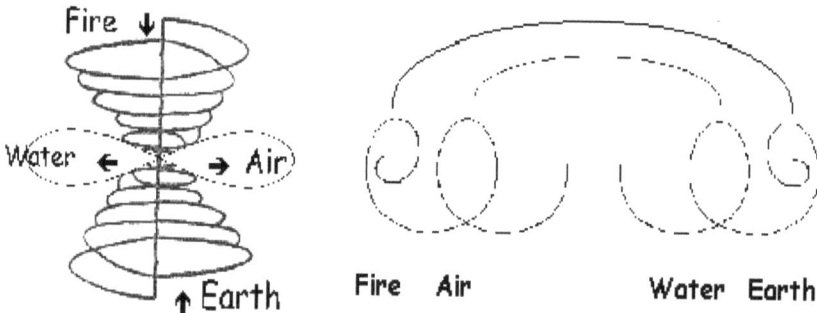

Fire   Air   Water   Earth

the horizontal plane is formed through the tension of Air and Water.

LEVEL FIVE - The Planets – 7-fold

Level Five is the plane of the planets. This is the sphere around the Earth starting at the Ionosphere and extending to the edge of our solar system. The order used is based upon the planets being placed according to the length of their orbital cycles as they circle the Sun. At present this pattern extends out to the orbit of the most recently discovered planet, Pluto. However when dealing with manifest life it is common to use only the 'traditional' planets we can see with our eyes: the Sun plus the six planets, which includes our Moon, Mercury, Venus, Mars, Jupiter and Saturn.

The outer three planets were consciously discovered in more recent history and are associated with areas of extra sensory perception and the collective unconsciousness. The 'original' seven planets, are connected to the physical manifesting streams of the universe. Saturn is seen as the limit or boundary-giver in life.

The natural expansion of the vortex spiral sees the elements of Air and Water from level four continue to divide forming a third polarity. At this level once again the planets are divided into several patterns. We can initially identify the same divisions we found at the 4-fold level: the macro and micro, or external and internal polarities.

Macro Polarities

As we are dealing with life on planet Earth, we observe the solar system as if we, the Earth, are at its centre. From this position, as we look out, we see there are three planets closer to the Sun than us, and three further away. These inner planets are the Moon, Mercury and Venus, while the outer planets are Mars, Jupiter and Saturn. The macro polarity of the solar system is immediately obvious from this image.

The outer planets are considered assertive and 'worldly'. Traditionally these three planets have been associated with the following attributes: Mars being actions taken in the world, Jupiter with the cultural and philosophic condition of any group, while Saturn governs the rules and bureaucracy of any social grouping.

♄	♃	♂	SUN	♀	☿	☽
Saturn	Jupiter	Mars		Venus	Mercury	Moon

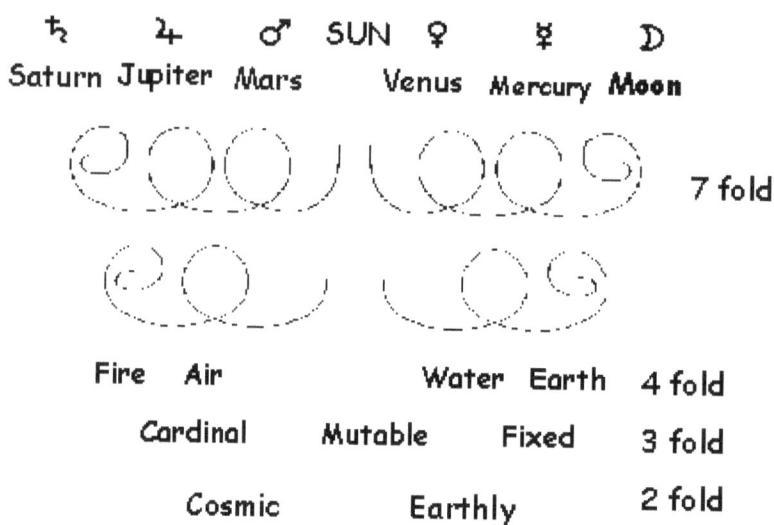

			7 fold
Fire	Air	Water Earth	4 fold
	Cardinal	Mutable Fixed	3 fold
	Cosmic	Earthly	2 fold

The inner planets are more personal and retiring in nature: Venus governs the receptive side of relating and what makes you feel good, Mercury generally covers communication, while the moon covers the personal emotional response, instincts and nurturing needs.

The outer planets are concerned with that which endures beyond the first cycle of any impulse, while the inner planets are associated with more transient phenomena. Each planet has its own individual quality, but also functions as an 'organ' in the body of the solar system. Indeed medical astrology ascribes each planet the rulership of specific organs in the human body.

Micro Polarities

The patterning of the micro polarities continues in the same way as we saw in earlier levels, as a development between an outer planet

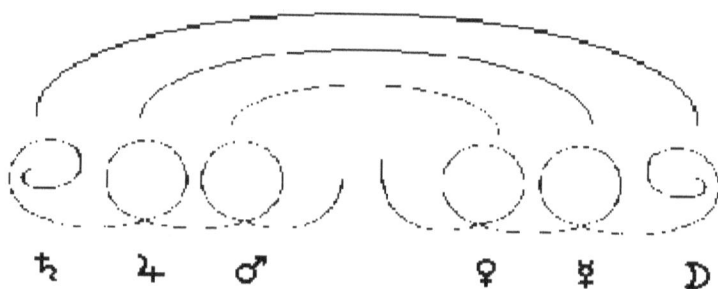

♄	♃	♂		♀	☿	☽

and an inner planet. These individual polarities form creative harmonies which can be identified in many different areas of life.

The Moon and Saturn.

The moon is the closest 'planet' to the earth, and is associated with the primary nurturing and uncontrolled development of life. If left to itself, the moon forces would create life as an amorphous mass existing only to multiply as in primary cell division,. Hence creating the concept of 'The Blob'. It is emotional and 'watery' in quality and associated with a primary stage of intuition, through unconscious reaction to sensation. The first few years of a baby's life, are very much moon years.

Saturn is the wise old man, with an authoritative image, who brings structure and form to the Moon's growth. It brings form, manifesting as the human skeleton and skin, responsible action and social systems, or bureaucracies. Left to itself it would continue until we and its other forms, become sclerotic stalactites. It is only through the balanced interplay of the Moon and Saturn (the child and the adult) that life finds form and maturity, at their allotted optimum time.

This polarity is - the will to manifest.

Jupiter and Mercury

Traditionally, both Jupiter and Mercury are considered to govern intelligence. Mercury is quicker in nature, and scurries about collecting all the facts. Jupiter is the philosopher who examines, balances and gleans wisdom from Mercury's information.

The image of Jupiter is of the well-rounded mature entity, governing the cultural life of a community and the middle years of life from 30 till 50 years. This is the period when one has accumulated experience and wisdom and still has the health to act consciously. It shows in plants as well as humans. It governs the deciduous trees that often form well rounded spheres, for instance the English Oak, and fruit trees of all kinds.

Mercury on the other hand governs climbing and creeping plants that need a supporting tree to hold them upright. The adolescent with their endless stream of unfocused energy personifies Mercury. The key thought here is that both planets expand whatever they touch.

These two can be seen as - the thinking planetary polarity.

Mars and Venus

This is the polarity of the feelings.

Venus epitomises all that is beautiful and harmonious in the environment. It can be said that its sole aim is pleasure, and being an inner planet, it gains pleasure through attraction i.e. by its beauty it attracts. It is concerned with an easy interchange between any two objects and aims to achieve harmony. It creates the milieu for social interaction to occur. Romance and foreplay are both Venus' environments.

Mars, on the other hand, takes action to get what it wants, and is prepared to achieve its aims at any cost. With Mars there is always some loss in the process of gaining the prize. It is gross compared to Venus but in their intermingling a harmony is reached and Mars happily gets for Venus anything she wants.

In so doing these are the planets of relationship and feeling.

The Sun

The Sun is symbolically the central energetic individual, the "I AM" presence, who mediates and harmonizes the creative tension set into play through the planets. As such it is a balance of all the above energies.

It is interesting to note that there are three planets on either side of the polarity. One would expect level 3 to be functioning here also. While the exact associations of the planets to the modes may not seem apparent at this stage, the associations will become clearer once we look at the Zodiac.

Can you anticipate which of the three modes relates to the planets on each side of the polarity?

While these two arrangements might be considered the archetypal structures of the planets there are a few other patterns of the planets used in Astrology which are explained in the Chapter: The Planets

The "far-out" Planets

Uranus, Neptune and Pluto, being collective influences generally work in areas beyond the individual, in group phenomena. Uranus governs Group Ideologies, Neptune - Group Faiths while Pluto governs Group Movements. If an individual has personal planets closely related to these outer planets it gives them the opportunity to connect with collective impulses more closely. Telepathy and psychic perception arises from this situation. Uranus creates the Occultist, Neptune the Mystic and Pluto the Shaman.

They do indeed create 'far-out' individuals, some might even say " beyond the fringe".

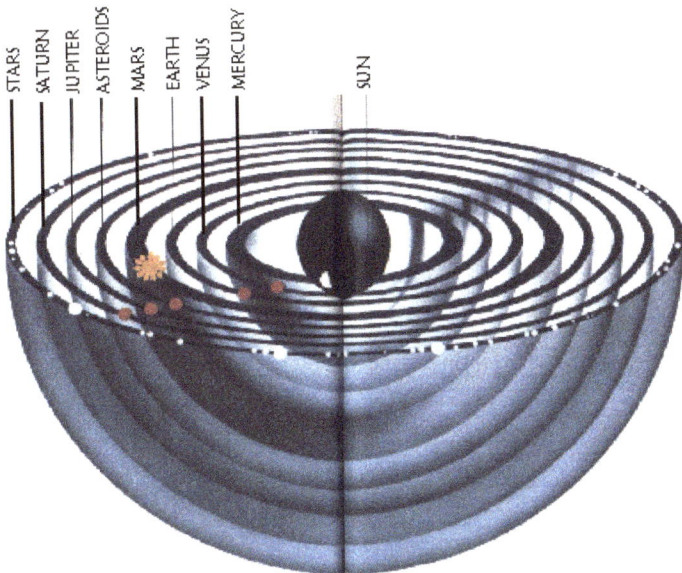

The Solar System Onion

LEVEL SIX - THE ZODIAC - 12 fold

This level of the vortex spiral relates to the galactic level of our existence. Our solar system is part of a greater stellar body called the Milky Way galaxy. This is a spiral cluster of around 100 billion stars which is approximately 100,000 light years in diameter. Our Sun is positioned about 30,000 light years from its centre, on one of the outer spiral arms. All the stars we see with our naked eye are within our own galaxy.

The zodiacal constellations are those constellations of stars at the back of the path the planets mark out across the sky. This path is called the ecliptic. As we saw earlier, manifestation in an organised gyroscopic system, occurs on the horizontal plane. It would therefore appear that everything in our Galaxy and Solar System follows a similar pattern, within its given system. Even though the planets, for instance, are at different distances from the Sun, the planets all revolve around the Sun within approximately the same 5 degree arc of the sky. The fixed star constellations in the background of this path, form what is called the Sidereal Zodiac. The other fixed star constellations which occur all over the sky, as seen from the Earth, are not however part of the Zodiac. The horizontal plane of the Galaxy is the groupings of stars called the Milky Way.

Equal and Unequal divisions of the Zodiac

Until the time of Christ all cultures used the CONSTELLATIONS around the ecliptic as their zodiac. Since shortly after Christ (around 200 AD) however our western cultures have developed a zodiac based on a division of the Sun's annual orbit around the Earth, starting at the spring equinox of the northern hemisphere. Rather than being based on the constellations, which form the ecliptic background, this Tropical Zodiac starts each year when the Sun is at the vernal equinox. An equal twelve-fold division of the Sun's path is marked off from this point, thus marking off the SIGNS of the zodiac. At the time this system was re-established by Ptolemy, the Vernal Equinox was at 0 degrees of the constellation of Aries. So it was a natural enough assumption to create an equal twelve-fold division alongside an unequal twelve-fold division. It is important to note the Tropical SIGN

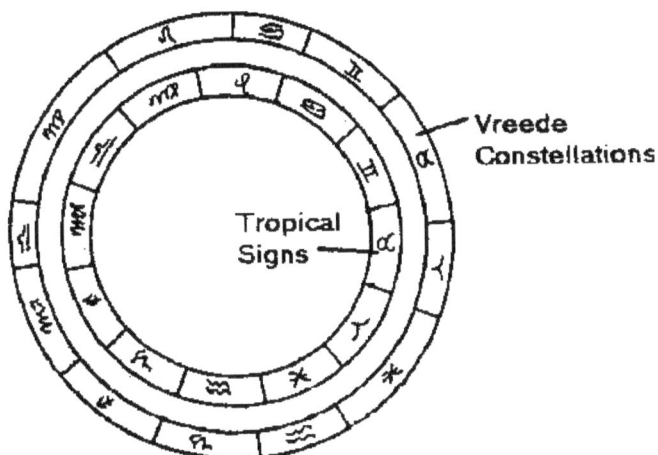

The Two Zodiacs

Zodiac is a twelve- fold division of a planetary sphere – (The Sun) anchored to the Earth, while the Sidereal Zodiac is a twelve-fold division of the Fixed Stars based upon the axis of the fixed stars Aldebaran and Antares.

Due to the Vernal Equinox moving backwards through the constellations one degree every seventy-two years, the start of the Tropical Zodiac is now at the third degree of the constellation of Pisces, each year. This has created a separation between the two zodiacs so that the moon's path through a particular constellation now occurs two days before its journey through the corresponding Tropical sign.

At this stage let it suffice to say that both are right and that they have different applications. A more thorough investigation of this question is presented in later chapters.

Internal Relationships within the Zodiac

The same rules and patterning's we have seen in the previous layers of the vortex spiral all apply at this level. This is one of the most striking facts which acts as certain proof that the Astrological Spiral is indeed a Universal Theorem. All earlier patterning's are expressed in subsequent levels, while the last level includes all earlier patterning's. This is indeed a wholistic formula, built out of the bones of outer creation to describe the inner workings of creation. So much so, that it is easier to understand the constellations' inner qualities (or Signs) by seeing them as a sum of their parts rather than by learning the qualities of each one individually. The influence of each zodiac constellation is made up of one part of each of the previous layers, namely it is ruled by 1 polarity, 1 mode, 1 element and 1 planet.

Two-fold in the Zodiac

At the twelve-fold level we see the divisions made by the planets forming the primary basis of the patterning however each planetary piece now internally polarises to give us a positive and negative side of the planetary impulse. As an image of the two- fold level carried to this higher level, one side is governed by the Earthly Female pole while the other are of the Cosmic Male pole.

This polarising of the planetary influence provides us with a dual planetary reference. We now have a primary or Cosmic planetary

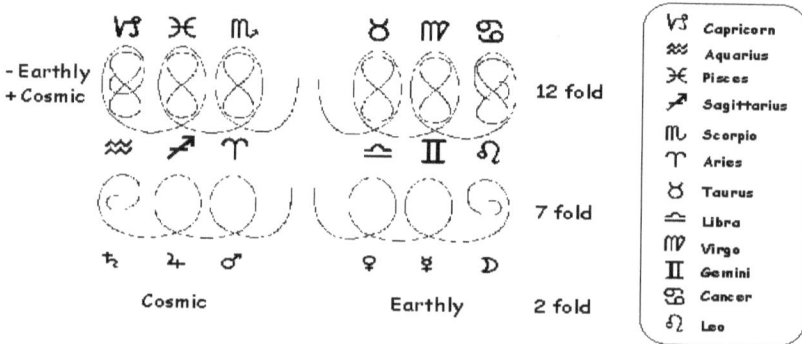

process and a secondary Earthly planetary process. As well as this we can include the macro planetary division of the Outer and Inner planets. This provides a fourfold picture of Cosmic Outer planets, Earthly Outer Planets, Cosmic Inner planets and Earthly Inner planets.

This is the first indication of the order offered by Goethe and Lievegoed. Goethe indicated the incarnating 'Being' and excarnating 'Manifestation process while Lievegoed associated this to the incarnating and excarnating planetary processes.

This diagram shows how this is imaged in the zodiac relationships as well, hence we have a reference for the primary planets being a

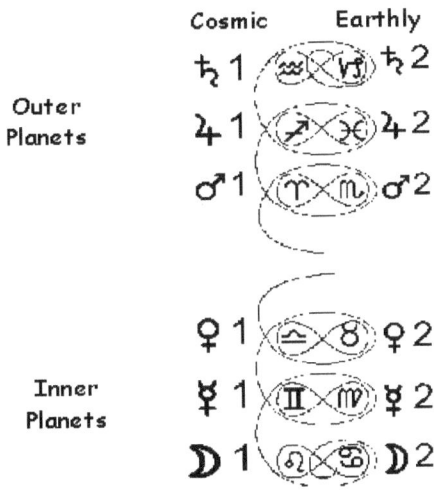

	Cosmic	Earthly	

Outer Planets

$\hbar\,1$ (≈ ♑) $\hbar\,2$

$4\,1$ (♐ ♓) $4\,2$

$♂1$ (♈ ♏) $♂2$

related to the cosmic stream and the secondary planets being a reference for the earthly stream of the zodiac.

Inner Planets

$♀1$ (♎ ♉) $♀2$

$☿1$ (♊ ♍) $☿2$

$☽1$ (♌ ♋) $☽2$

This process of relationships can continue for the other layers of the vortex in regard to the zodiacal constellations.

3 fold Modes & 4 fold Elements in the Zodiac

A wealth of association develops from this.

The Law of polarity divides the zodiac into two groups. In the Positive Cosmic group all constellations are either of the Fire and Air element. In the Negative Earthly group all the constellations are of the Water and Earth elements.

The macro polarities operating earlier between Fire and Air or Earth and Water are always in polarity to one another at the twelve-fold level.

The Modes structure of Cardinal, Mutable and Fixed show up and further identify the four sub-groups within the zodiacal diagram. Each organise themselves into Cardinal

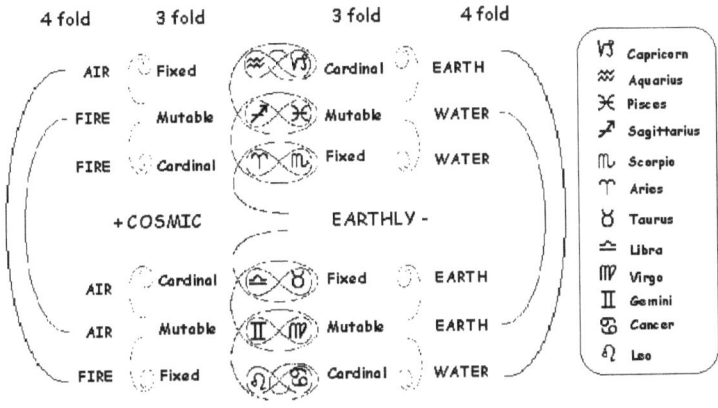

4 fold	3 fold		3 fold	4 fold	
AIR	Fixed	(≈ ♑)	Cardinal	EARTH	
FIRE	Mutable	(♐ ♓)	Mutable	WATER	
FIRE	Cardinal	(♈ ♏)	Fixed	WATER	
	+COSMIC		EARTHLY -		
AIR	Cardinal	(♎ ♉)	Fixed	EARTH	
AIR	Mutable	(♊ ♍)	Mutable	EARTH	
FIRE	Fixed	(♌ ♋)	Cardinal	WATER	

♑ Capricorn
♒ Aquarius
♓ Pisces
♐ Sagittarius
♏ Scorpio
♈ Aries
♉ Taurus
♎ Libra
♍ Virgo
♊ Gemini
♋ Cancer
♌ Leo

Mutable Fixed structures, true to Level 3. This four-fold process suggests the gyroscopic form and so from this cross-reference it is possible to gain an idea   as to the 'identity' of each of these four sectors of the zodiacal gyroscope.

Sign & Constellation Qualities

To determine the characteristics of any particular Sign/Constellation, the individual parts are pieced together.

Take Aries for example.

From the chart it can be seen that it is a positive, cardinal, fire sign ruled by the planet Mars. This indicates that Aries is an expansive, leading, optimistic sign that is prone to bursts of assertive action in the achievement of its goals.

Scorpio on the other hand is also ruled by Mars, however the polarity is negative, the mode, fixed, and the element– is water. It could be said that Scorpio takes assertive action to conserve and maintain the support structures it has attracted to itself.

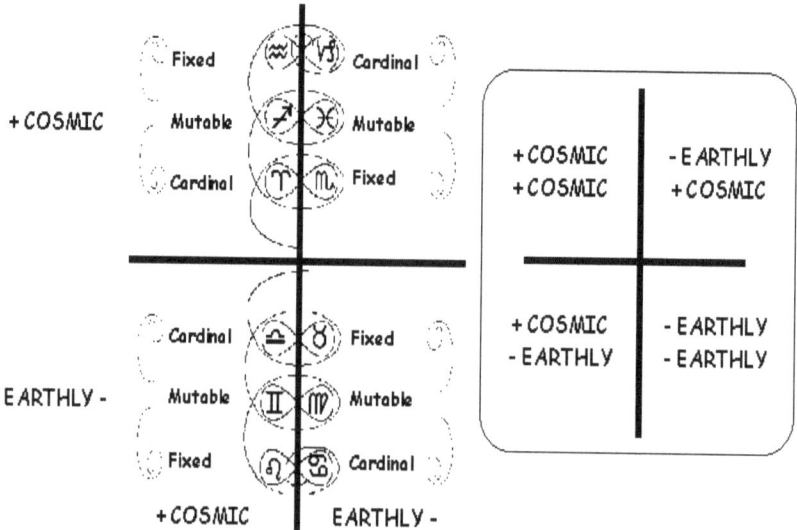

Conclusion

All the above diagrams show the inherent harmony of the zodiac. All parts of the twelve-fold (level six) relate to each other according to the 'laws' indicated at previous levels. In this way the conclusion can be drawn:

For such order to be expressed in this final stage,

all earlier relationships and conclusions

about the theorem must also be correct.

It is this essential harmonic phenomena, that all the parts make the whole and the whole reflects the parts, that most fascinates me about this approach to Astrology. It indicates Astrology has the potential to be truly holographic in its nature and application to life. This very phenomenon of implicate order, may well have been the downfall of Kepler, the famous Astrologer/Astronomer, who held that the paths of the planets must be perfectly circular and not elliptical. He possibly projected this philosophic harmony of

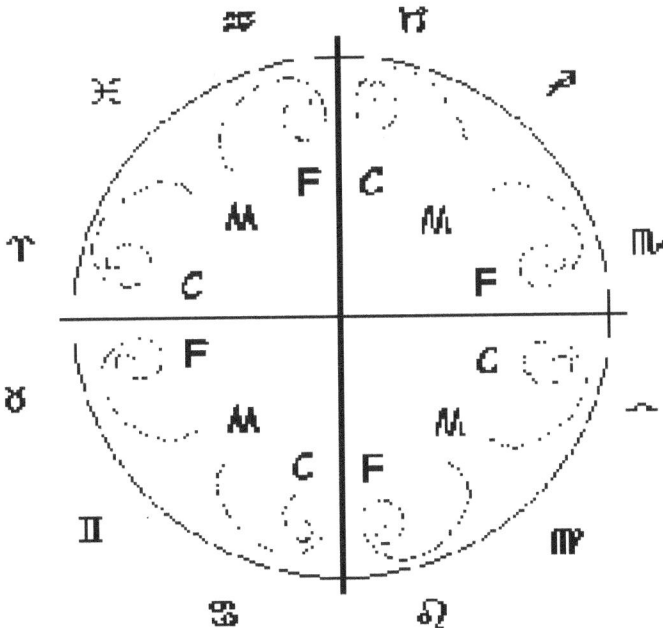

Astrology onto his Astronomical observations of our complex living organism, the solar system.

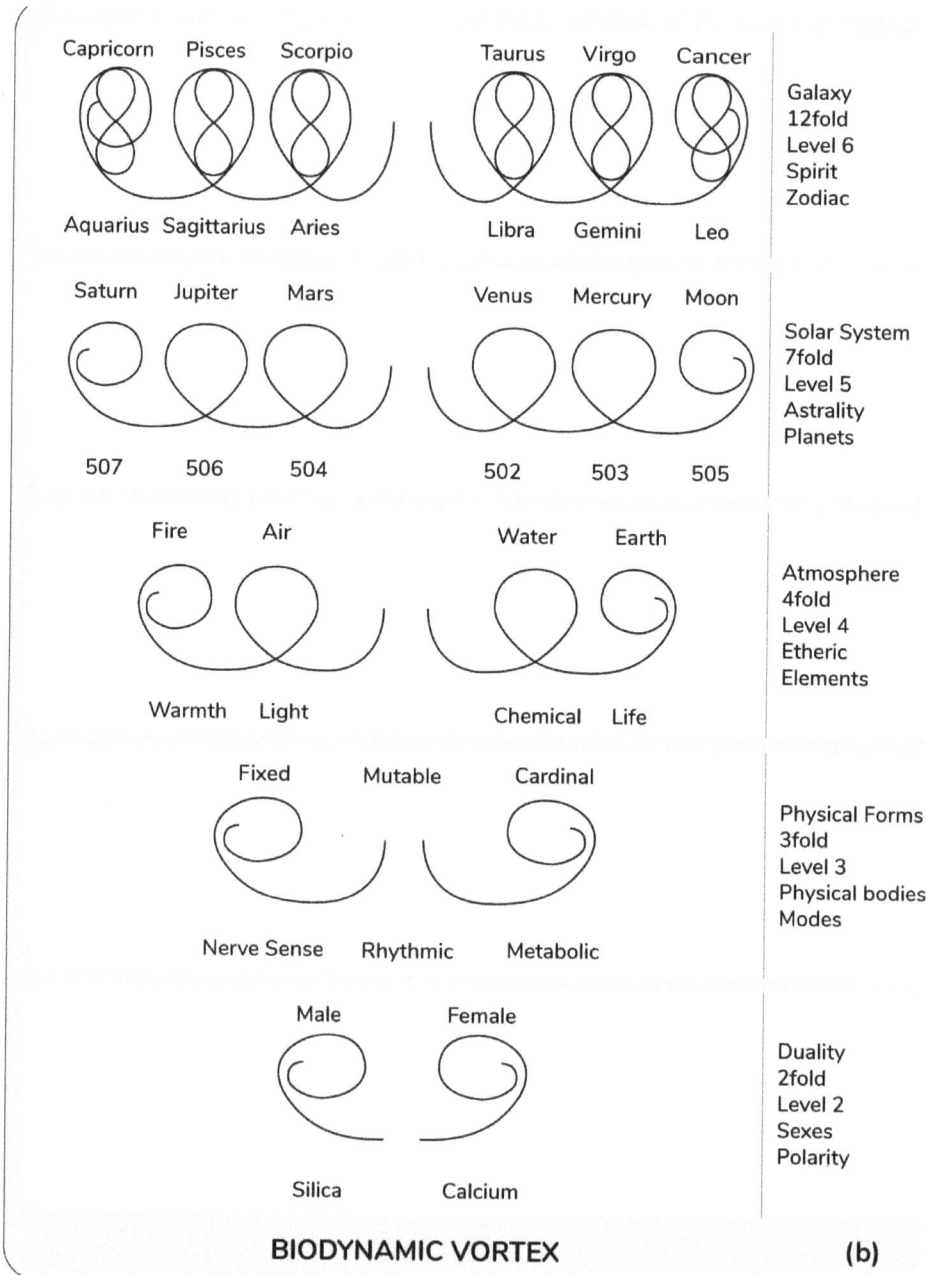

Capricorn Pisces Scorpio	Taurus Virgo Cancer
Aquarius Sagittarius Aries	Libra Gemini Leo

Galaxy
12fold
Level 6
Spirit
Zodiac

Saturn Jupiter Mars	Venus Mercury Moon
507 506 504	502 503 505

Solar System
7fold
Level 5
Astrality
Planets

Fire Air	Water Earth
Warmth Light	Chemical Life

Atmosphere
4fold
Level 4
Etheric
Elements

Fixed   Mutable   Cardinal

Nerve Sense Rhythmic Metabolic

Physical Forms
3fold
Level 3
Physical bodies
Modes

Male   Female

Silica   Calcium

Duality
2fold
Level 2
Sexes
Polarity

BIODYNAMIC VORTEX   (b)

EVOLUTION ON A PINHEAD

A broad but simple definition of Bio-Dynamics is that it is an agricultural system based on sound organic practices that consider the immediate environment, and that all life on earth is a result of the forces streaming onto our planet from the Fixed Stars and their planets .

In an attempt to grow produce of optimal nutritional quality, Bio-Dynamics challenges the practitioner to view the environment as a wholistic system of life and each farming enterprise as an individuality in its own right.

Some further definitions;

'Sound organic practice' primarily means that soil fertility is maintained through composting, green manuring, crop rotation and diversification (with inclusion of animal manure at some stage of the process). Also, that any remedial measures which may be necessary at any stage of production, are sourced from living materials and are not harmful to the environment in part or whole.

The "environment" is understood in the widest possible sense. It is not only the immediate soil in which plants grow, but the landscape in general, the planet as a whole and the solar system in which this planet is moving. One is left with, image of being a cosmic citizen just because one wants to grow cabbage!

It cannot be hoped to outline all the intricacies of Steiner's philosophy, however a short outline of some sections of it is necessary in order to work with the Bio-Dynamic understanding at some depth. For a deeper understanding of these concepts refer to the reading list.

In his eight agricultural lectures, Rudolf Steiner outlined a complex picture of our planet. He described it as a energetic entity that sustains life, through its interaction with the creative energies and rhythms within the universe. This is a broad picture that covers cosmic streams of energy, planetary bodies, physical and subtle bodies of the earth, plants, animals and man, and the chemical elements, particularly those of protein.

The most practical outcome of the course of lectures is a series of preparations used to help harmonise all these forces he describes. These preparations, and the overall insights he gave into nature, are truly a gift to humankind. The preparations potentially give humankind the conscious control we have so long desired over plant growth, and for which we have had to resort to artificial and poisonous chemicals to achieve. These preparations help to balance the interplay between the physical and more subtle bodies of creation.

Before we go into the preparations, we need to understand Steiner's view of the plants, animals and humans.

To answer the questions being asked in this book, we need to quickly explore some aspects of RS's world view. Most notably the structure and physical workings of the energetic bodies. This is the crux of this whole discussion.

In line with most energetic teachings, Rudolf Steiner outlined life on this planet as manifesting due to an interplay of energetic bodies' and physical elements. These bodies have slowly 'evolved' through aeons of time.

He outlined in 'Occult Science - An Outline' (4), that creation has developed so far over a series of four great Eras. These Eras, which he called the Ancient Saturn, Ancient Sun, Ancient Moon and our present Earth evolution, equate to the 4 Yugas of evolution, outlined in the Hindu scriptures. These being the Swarpa yuga, The Dwarpa yuga, the Tretra yuga and the Kali yuga. Steiner suggests there are three more of these great ages to follow our present age, while eastern belief is that the Kali yuga, our present era is the last era of this cycle.

Within these great Eras there are sub cycles we can identify. There are three time scales to consider. There are the extremely long cycle mentioned above, the Eras. A secondary scale called the Ages. There are seven ages within each of the above long Eras.

Within each of these secondary ages there are seven internal cycles called the Cultural Epochs, of approximately 2165 years each in length. This time period corresponds to the time it takes for the

Precession of the Equinox to move through on equal constellation of the Zodiac. Hence the constellations describe the quality of the cultural ages.

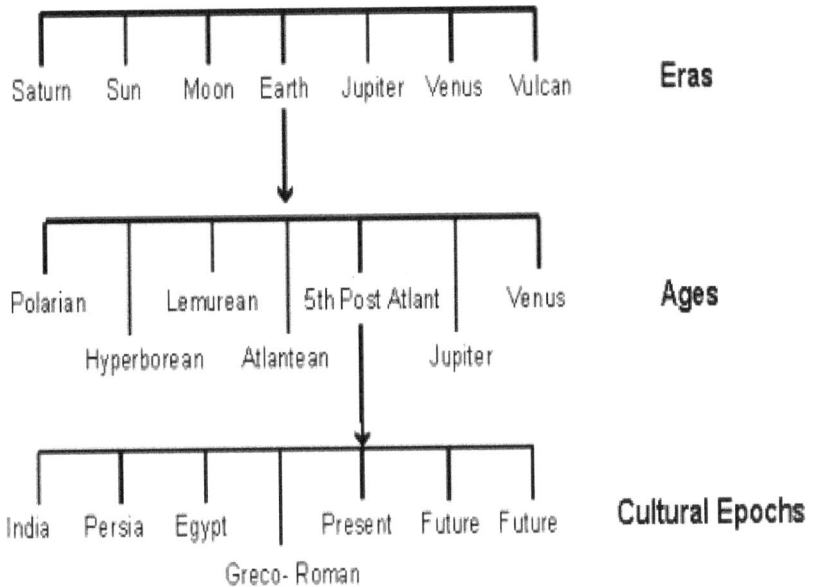

| Saturn | Sun | Moon | Earth | Jupiter | Venus | Vulcan | **Eras** |

| Polarian | Lemurean | 5th Post Atlant | Venus | **Ages** |
| Hyperborean | Atlantean | Jupiter | |

| India | Persia | Egypt | Present | Future | Future | **Cultural Epochs** |
| | | Greco- Roman | | |

We are presently in the European Cultural Epoch of the 5th Post Atlantean Age of the Earth evolution Era. This whole story is the evolution of our Solar System only.

During each of the large Eras of evolution new aspects of creation came into being. Most notably in each Era a new energetic faculty or 'body', has been added to life on this planet. Both on a microcosm level in life forms and on a macrocosm in our environment. (5) So, in this fourth Era of evolution, the Earth evolution, life now potentially consists of a Spirit or Ego. an Astral or sense body, an Etheric or life body and the last to manifest, in this most recent period, the physical body.

(As this is an essay using Anthroposophy as its base I have chosen to use the common Steiner term of the incarnated Spirit -The Ego.

The four physical structures within our Galaxy are seen as the 'homes' of the energetic bodies. The Galaxy, where the formative forces of the Fixed Stars originate is the Cosmic home of the Spirit. Science tells us it is only the electro magnetic impulses coming from these stars that fills space. Otherwise it is 99.5% Hydrogen in a vacuum.

The Solar system is the 'World' base of the Spirit Sun and World Astral and the home of the Planets . This Astral activity is responsible for stimulating sensations and psychological responses we have in response to energies coming from outside us in the World.. As we will discuss later it is primarily through studying the planets that we can observe the movements of our psychology.

The atmosphere has developed from the Earth. As the oxygen level has increased so it has been able to support an ever increasing diversity of life. This is the home of the 'World' Etheric or life body.

Naturally physical life takes place on the Earths surface.

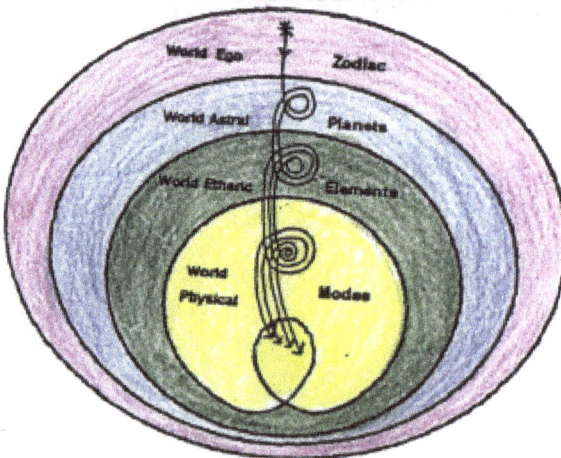

These 'bodies' while finding their origin in our macrocosmic environment, have been internalised to varying degrees by each of the different kingdoms of nature. Each kingdom 'personalises' the 'bodies' in their own way. Thus giving us the many different manifestations of the life remaining on our planet.

This story can be told in a slightly different order. RS tells the story of 'The Three Worlds'. He first talks of it in 'Theosophy', He says that once we have passed the veil of matter, we come upon an energetic world that has three separate energetic worlds, existing in the same place and time, however with completely separate workings and laws. He uses this as the basic energetic matrix standing behind manifestation. He calls these Worlds the Spirit World based in the Stars, the Soul / Astral world based in the Planets, and the Third

Manifest Word of the Earth. These are all real energetic things, so he is not asking us to believe anything .

The Fourth World of the Atmosphere, that supports the Etheric and thus Life, comes after the Earth has formed and cooled some. Life then produces more Life, and here we are.

The Lemniscate - Cosmic Bodies, Earthly Life

Before we look at each kingdom of nature it may be of value to look at the process of incarnation. In the kingdoms of nature we are looking at the cross over point between the macrocosmic formative structure of Life and the microcosmic manifestation of Life. There is a hint in the basic process of Life Cell division. A Cell is stable, it begins to pulsate and it then divides. It is here the lemniscate becomes a useful model. (Hubble photos) A lemniscate is the form created when a flat spherical plane is twisted in half. This in turn creates two surfaces or dimensions through a middle junction point, however the lines of connection between the two dimensions are never severed.

As Above So Below

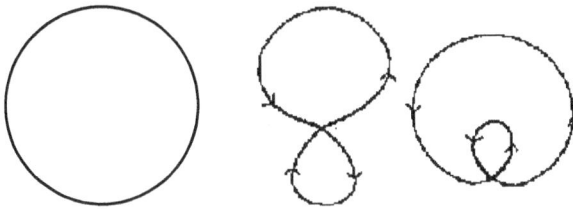

The spiralling forms of life, which are mirrored at every level of the cosmos, provides a starting point. A running river moves continually through spiralling forms, which twist and turn eventually folding in upon themselves. It firstly twists to form a lemniscate and then folds back under itself.

I imagine a spirit spark begins to spin in a centripetal motion slowly gathering speed and eventually creating a double ended spiral. Eventually one end of the spiral inverts in upon itself which in turn creates a lemniscate in space. The top of the lemniscate is Cosmic forces and 'matter', while the bottom starts to collect Terrestrial forces and matter in its inwards spiralling motion. This spirit is essentially becoming a black hole in space attracting matter and light to itself. This spiralling motion gathers its own momentum until it folds in upon itself, as a third stage, collecting more matter and forces, both Etheric and Physical into its physical body. It starts to cast a shadow in space. The womb of women can be seen to be such an enfolded internalised life forming space, open to world out of which life again proceeds.

The connections between the outer Cosmic Body and the internalised body are never severed. Hence as the macrocosm moves, so life moves.

Pssttt... message from 2025

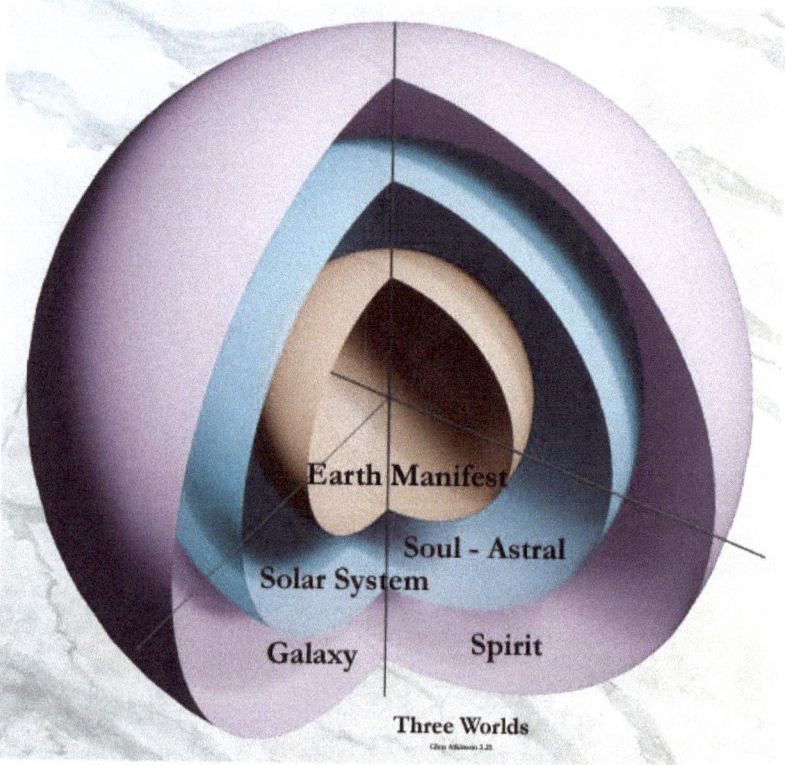

Earth Manifest

Soul - Astral

Solar System

Galaxy

Spirit

Three Worlds
Glen Atkinson 3.25

THE BIG PICTURE

I wish to thank Malcolm Gardner and Hugh Lovel for there parts in discussions on BDNOW from which this material has been developed.

Rudolf Steiner has outlined a multi layered reality for us to contemplate. The difficulty for the neophyte of his world view especially when reading his lectures as they randomly appear in ones life, is that he talks about these different layers or dimensions of reality as they relate to the subject at hand. Defining the overall BIG order in this manner is very difficult. RS outlined his basic world view in the books 'Theosophy' and 'Occult Science- An outline'. However much more is presented through his lectures.

I have found it useful to bring some clear understanding into the 'structure' of Rudolf Steiners view of creation. I do not wish to go into the process of creation, which is adequately covered in 'Occult Science'. I wish to outline the big picture of the 'parts' of creation we are confronted with when exploring life on Earth.

Energetic reality and material manifestation are so intimately linked that we can look to the Astronomical structure of creation, as we presently understand it, as the basis of any structured view. But first, what are we looking for? RS describes the overview of creation as a threefold being. The Three Worlds image can be found influencing many understandings. On the large scale, we have a personalised or internalised reality which we can see manifest in the four kingdoms of nature, A 'World' reality which is best seen as the neighbourhood in which the Earth lives namely the Solar system. The Galaxy in which we exist is the third 'Cosmic' reality. This also encompasses all stars, planets and galaxies beyond our own.

This order is the basis of the energetic bodies at the base of natures manifestations.

Cosmic spheres

To appreciate the cosmic spheres one needs to stretch ones imagination. A slow journey through an astronomical chart of creation is invaluable to achieve this. The National Geographic put one out some years ago. The charts I refer to start with the Earth and then macroscope out to the Solar system (which has a diameter of 11 light hours), to our local stars and then to the Galaxy, which has a diameter of 100,000 light years. After moving through our local group of Galaxies (5 million light years) it enlarges to our local supercluster (100 million light years) to come to the observable universe with a diameter of some 30 billion light years across. From here we have to proceed briskly through Hindu scripture which tells us that the superclusters of stars we can see - each around 100mill light years across- are only a small number of the amount which are breathed into existence by Lord Brahma (the four headed, first created being who sits upon a lotus flower growing from the navel of Vishu, who in turn is lying upon a serpent in the ocean of milk.)

It is with Lord Brahma that our story can start. All the universes breathed into creation by Brahma are 'swimming ' in a sea of chi or cosmic ether. This force is said to permeate all of creation and when it comes into manifestation appears as a growth process. In discussions with Hugh Lovel on the BDNOW email list (BDNOW@igg.com), Hugh has assured us this **cosmic ether** comes from beyond the stars and flows towards our galaxy and enters each subsequent gyroscopic being - galaxy, Solar system or planet - through the horizontal axis. So with the Earth, it enters our atmosphere at the equator and travels towards the Earth before being sucked out the poles and off to the Sun. More about this later. This is the Cosmic ether, which is 'All that comes from Above' and not to be confused with the World Etheric which is organised by the Earth.

There are quotes from Steiner which state the Astral forces come from the stars of the Zodiac, while the Egoic forces originate in the stars beyond the Galaxy. The forces described here are the Cosmic Astral and Egoic forces which develop in other solar systems and radiate towards our Solar system from the periphery.

It may help to understand that Astral forces are a result of the activity of the Ego or Spirit. They do not develop by themselves. Planets develop from the activity of a Star. Lets use the human as another example. The spirit of an individual wishes to incarnate into life. It joins together with a physical form which brings with it the development of an etheric body. Once the spirit is 'in life' it begins to have experiences from all its surroundings. The experience, sense impressions, memories and emotions created by any activity becomes accumulated to form the Astral body. This Astral reservoir , over time becomes so full that it begins to appear as a personality. It also starts to be seen through the forming influence it has over the etheric/physical organism. At this stage our personality can be seen in our bodily form.

On a Cosmic scale this same process is seen in the formation of stars and planets. Just as the Spirit is the central organising influence of a life form so we see a Sun is the central being of a Solar system. It is out of the whirling vortexes of Hydrogen that eventually a Sun ignites and begins to radiate more energy than it absorbs. From this gyroscopic generation and movement the Sun develops a large series of spherical electro magnetic fields. Along the horizontal plan at the equator of this 'being' matter begins to accumulate, due to the electromagnetic resonance of the whole being. At rhythmical distances along this equator the formation of planets occurs. Eventually the planets accumulate most of the matter in its particular sphere and becomes a solid form. Due to the formation of this mass it then begins to exert gravitational and electro magnetic altering influences upon the whole system.

This macrocosmic story is the same as that seen in the formation of an individuals Astral body. It has long been understood that the Spirit and Suns have an affinity and this gains more relevance when we appreciate Rudolf Steiner's insight that Hydrogen is the physical carrier of the spirit. Now it is a simple enough jump to follow the metaphor and see that the Astral formation and that of the planets are similar. From my studies of Astrology I have come to see that the Sun in a birth chart is the residence of the spirit, while the planets of the birth chart and the movements of the planets in the sky act as excellent indicators of the activity of the Astral body in a humans life. From this I am in no doubt that the Astral activity is related to planetary activity while the Ego/Spirit is created by Stars. The Spirit is actual Star Forces. No woo woo belief needed.

To go back to RS earlier comment about Ego and Astral formative force originating in the zodiac and beyond, we can see he is talking of Cosmic Egoic and Astral forces. All stars will generate Egoic formative forces and their planets will focalise or create Astral forces. RS says the Astral forces are sourced from the region of the zodiac, while the Egoic forces come from beyond our galaxy. The stars of the zodiac and particularly the ones we can see are in our galaxy. So RS is indicating the Astral forces developed by the planets of the other stars in our galaxy, radiate towards us, and have an influence upon the formation of both the World and individual Astral bodies. The Egoic forces are much stronger and forces from all other stars have the power to enter into our galactic 'egg' and act as formative forces as well.

This then forms the basis for the Cosmic spheres of our existence

Cosmic Ego is sourced in the Stars activity.

Cosmic Astrality is sourced from the planets around the stars of our galaxy and most notably from around the stars of the zodiac

Cosmic Ether - all that comes from Above - flows from the periphery of creation towards the Earth entering at the equator.

World spheres

The World spheres are best seen as the 'energetic' bodies of our Earth. From the above examples it is clear the Sun at the centre of our Solar system is our central spirit being. Thus becoming the World Egoic source with the world Astral planets circling about it. The Sun is responsible for the formation of the planets, of which we are one. This indicates our Earth , as a planet, in itself is an Astral formation, however for us it holds a special more intimate significance, as we exist upon it. This firstly brings us into relationship with the physical body of the Earth, upon which we walk and then the life body of the Earth which is made possible for us to exist in due to the unique formation of the Atmosphere. It is within the Atmosphere that we find the right concentration of oxygen and the existence of light and warmth in such proportions that life can come into and be maintained in existence.

From this we can conclude the World spheres are

World Egoic forces emanate from the Sun

World Astral forces exist with in the planetary spheres

World etheric forces are maintained within the Atmosphere

World physical body is the Earth itself.

There is one development RS has made to this 'basic' picture. He has suggested that with the crucifixion of the Christ being 2000 years ago, the egoic forces of the Sun became more intimately connected with the Earth and that now the World Egoic forces work on us from within the Earth. As for the reality of this I can not say. No matter what energetic event has occurred we must still deal with the astronomical reality, that our Sun is the basic source of light, warmth and the immense sea of electromagnetic and nuclear forces we exist within. While using this picture makes it easier to understand how the egoic forces manifest as the Cosmic Silica stream of Earth energy pushing up from the Earth, we have also been given the physical mechanism for this process by RS. He outlined how the quartz crystals within the Earth act as a condenser, during the winter of the Cosmic and World Egoic forces streaming onto the Earth, (carried in by the Terrestrial Silica during summer) and then, due to the presence of clay in the soil, these Egoic forces radiate back out to the Cosmos through the stem and seeding formations of plant growth.

Personal Bodies.

The personal bodies of creation only find their full expression in Humans. We internalise the Ego, Astral and Etheric activities within our physical form. The lower kingdoms of nature only internalise different combinations of these bodies. The animal kingdom is a manifestation of Physical Etheric and Astral combinations, while the plant kingdom is generally only internalising the etheric within its physical form, while the mineral kingdom has just matter. When a body is not internalised it works with that kingdom from outside. Their 'Being' does not end at their skin, it extends into their environment, as a collective 'organ'.

Humans internalise all bodies

Animals are internalised Astral, Etheric and physical with the World and Cosmic Egoic forces working directly from the Sun and stars.

Plants internalise the etheric forces and have the Cosmic and World Astral forces working onto them from outside

Minerals have all forces working onto them from outside.

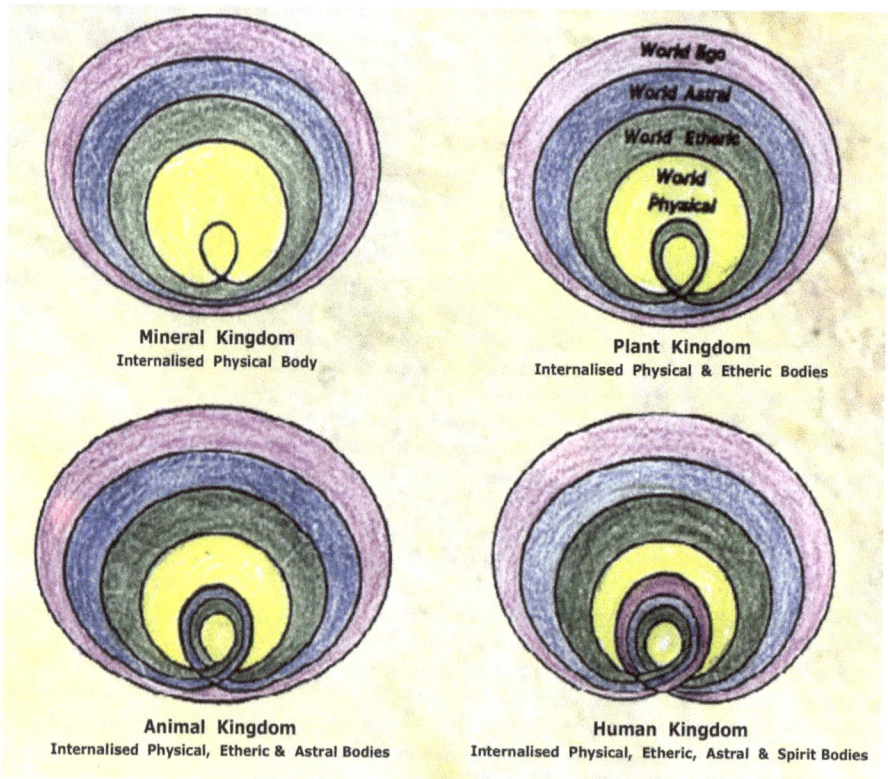

Mineral Kingdom
Internalised Physical Body

Plant Kingdom
Internalised Physical & Etheric Bodies

Animal Kingdom
Internalised Physical, Etheric & Astral Bodies

Human Kingdom
Internalised Physical, Etheric, Astral & Spirit Bodies

CREATION'S LAYER CAKE

Throughout my Biodynamic studies, I have found it a constant challenge to merge the myriad of ideas and `pictures' I have been presented, with into a cohesive realistic picture. Hence I have continually endeavoured to find the practical or reality based carrier of any concept. One result of this search has been the development of the diagram on the next page. In it, I am attempting to base the `idea' of the energetic bodies, presented to us constantly through Rudolf Steiner's work in their physical spheres or `homes' and then explore how each sphere or layer is connected to the others.

The energetic bodies, presented as the basic vehicles of manifestation by Steiner, may be a new concept to some people. They are however an important `reality' to understand and ultimately live with if one is to fully grasp what is presented to us in `Agriculture'. The closest understanding of the energetic bodies we have in our twentieth century western Christian tradition, is the Body and Soul. These two groupings can each be divided into two. The Body can be separated into its purely physical carbon based body and a `body' which carries the living growth aspect of our physical body. Many traditions call this the Etheric body. It is this Etheric body which separates from the physical upon death, which then sees the physical body disintegrate. The Christian Soul can be divided into two parts as well. There is the eternal personal spirit where consciousness resides, called by Steiner the Ego, and the Astral body which is mostly experienced and understood as our

CHRISTIAN	BODY		SOUL	
STEINER	Physical	Etheric	Astral	Spirit

SPIRIT *Galaxy*	♌ ♍ ♎ ♈ ♓ ♒		
	♋ ♊ ♉ ♏ ♐ ♑	Spiritlands	
	☉ ↓ ⚶	Collective Conscious	
	♅ ♆ ♇	Collective Unconscious	
Solar System **ASTRAL**	♄ ♃ ♂	Outer planets Personal Conscious	
	☽ ☿ ♀	Inner planets Personal Unconsciou	
ETHERIC *Atmo sphere*	Warmth Light Chemical Life	Ionosphere Ethers	
	Fire Air Water Earth	Stratosphere Elements	
	Hydrogen Nitrogen Oxygen Carbon	Troposphere	
	CHO Proteins Sugars 'Life'	Biosphere	
PHYSICAL **Earth**	Mg, Na, K, Al, Mo Fe, Cu, Ag, Au, Pb	Chemicals	
	Ca. Si.	Solid Earth	

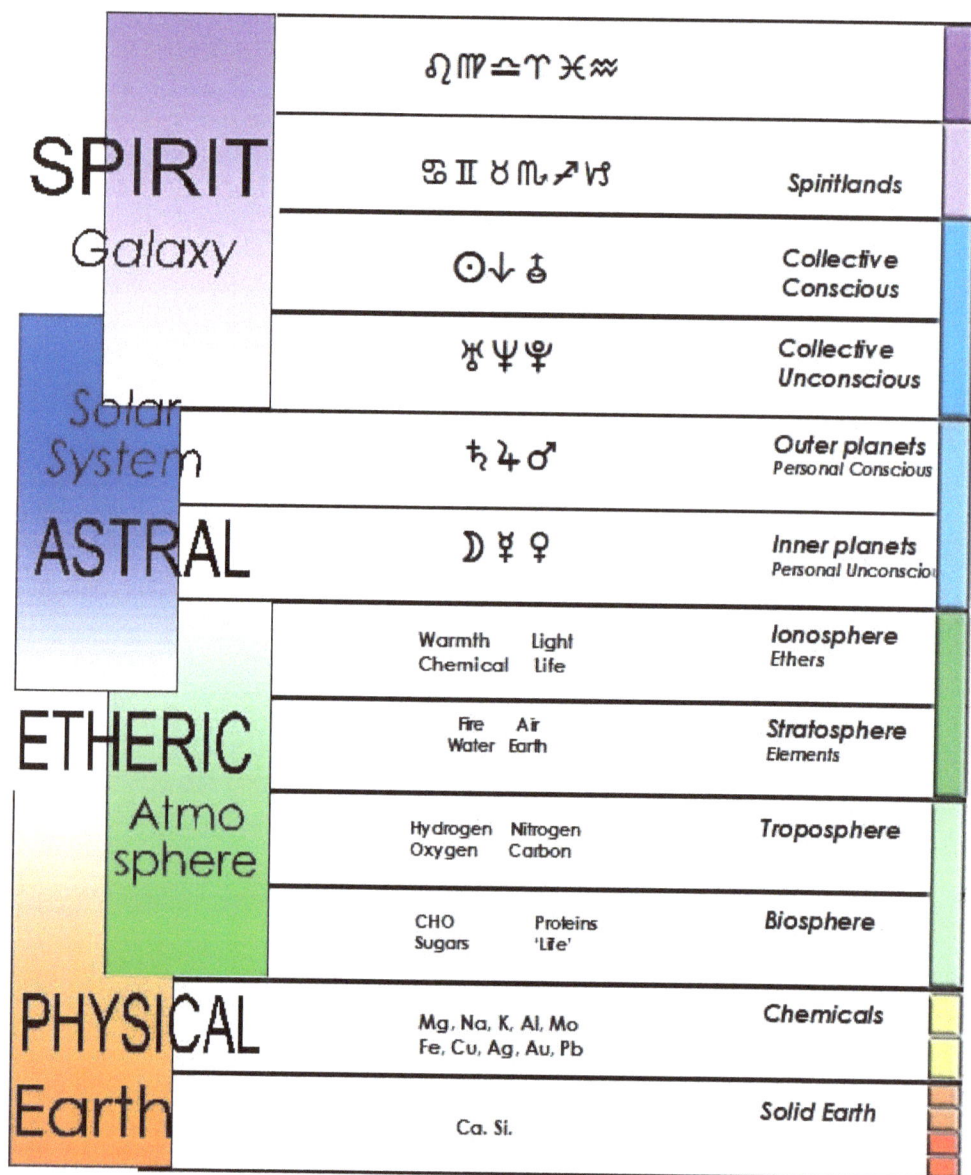

The layers of Creation

☽ Moon	♅ Uranus	♈ Aries	♎ Libra
☿ Mercury	♆ Neptune	♉ Taurus	♏ Scorpio
♀ Venus	♇ Pluto	♊ Gemini	♐ Sagittarius
♂ Mars	⚶ Persephone	♋ Cancer	♑ Capricorn
♃ Jupiter	↓ Vulcan	♌ Leo	♒ Aquarius
♄ Saturn	☉ Sun	♍ Virgo	♓ Pisces

psychology, emotions and psychic experiences. (I understand in Christian theologies the Spirit is reserved for `God the father' or the `holy ghost'. I would suggest in Biodynamics this is the Cosmic or Macrocosmic Spirit.) The difference between the Astral body and the Spirit / Ego is the difference between the Sun and its planets. The Sun is the generating vortex at the centre of our Solar system. It is out of the Sun's activity that the planets come into being. The same is true of the Astral body. It is out of the spirits active participation through life that our Astral body - sense impressions, psychological responses, neurosis - are formed. It can be said to `orbit' the Ego.

In Biodynamics, the sayings `As above so below' and `Life is a microcosmic image of the Macrocosm' takes on the most practical expression I have met so far. We are presented with the picture that these four bodies find their source and `homes' in the macrocosm. Life as expressed in the four kingdoms of nature, - the mineral, plant , animal and human - are bought into form and held there due to an internalisation or personalisation of small parts of these macrocosmic spheres. So we take for our selves a little piece of the Earth's matter, a dash of Atmosphere, a hearty slice of the Solar system and a pinch of Galaxy to become who we are. The proportions of this mixture are different for each kingdom of nature and within each kingdom each individual establishes a very personal balance as well.

In accordance with the picture presented by Lievegeod in " The workings of the planets in the life processes of Man & Earth" I have placed the Cosmic & World Spirit sphere with the Galaxy; The World Astral sphere with its physical carrier, the Solar System; The World Etheric sphere in its home, the Earth's Atmosphere, with the Earth naturally carrying the World physical body.

In nature there is no black and white. Any living process will intermingle at some point with its neighbour. So wherever these `physical' dimension meet, they crossover enabling an intermediate `space' of reality to come into existence.

Where the Etheric body enters into the physical sphere we have the development of the Biosphere. The minerals of the earth are lifted up by life processes - the Etheric body - into organic processes

69

of Carbohydrate, Sugars and Protein formations. Similarly where the Solar System and the Earth's Atmosphere meet we have the development of the electrically charged Ionosphere. Konig (7, Pg 89) shows how the Etheric ethers are manifest in bands of the Ionosphere. It is commonly accepted the elements of Fire Air Water and Earth are present predominantly in the Stratosphere, while the Gases of Hydrogen, Nitrogen Oxygen and Carbon (dioxide) exist in the proportions suitable for life only in the Troposphere. Within the Bio-sphere, these same four elements are the basis for organic chemistry. In this example we can see how the fourfold processes of the Etheric body organise themselves as different manifestations in each of these `etheric' layers.

Another point to observe is how polarities seem to develop between one layer and the next. With the elements, the Earth is the heaviest with the water lying on top of the Earth. The Air floats above the water while warmth rises above the Air. In Konigs example, the ether layers of the Ionosphere are described in the opposite order; the Warmth layer being placed at 45kms, the Light from 100km, Chemistry (ether from 200km and the Life (ether) from 300kms. This same process of inversion through a twisted lemniscate between, `sister elements' can be found in other layers as well and can be explored further later. The polarity relationships apparent between the inner and outer planets used in Biodynamics, especially when understanding the preparations relationships to the planets, is another example of creative polarity in action. (Lievegeod –8)

This diagram and others I hope to present, are for your deliberation, and are only a suggestion of what could possibly be. I suggest you place them on a wall and have them as a continual reference as you grow in your understanding of Biodynamics. Little pieces continue to fall into place, piece by piece over time. A book of pictures 'Organisation of Life' is available for free at the Glenopathy website.

In my experience Biodynamics is a physical science. It's concepts are based on `physical' realities and can be understood in a sequential common sense manner without the need to rely on faith, superstition or adhere to some abstract philosophy. It is a natural development out of sound Astronomical, physical scientific facts combined with an open lateral mind and imagination.

As Biodynamic growers we are attempting to consciously use as many layers of 'Creations Layer Cake' Diagram as possible, at one time. You will note our `conventional' cousins only use the first two.

There are several patterns we can draw on from macrocosmic creation in our attempts to understand our microcosmic environment. One I am finding of great use is the gyroscope. Creation is made up of a series of gyroscopic spinning `beings', whose matter manifests on horizontal planes, while spinning around a vertical axis supported by two vortexes. In fact Astronomy tells us space `matter' is sucked down these vortexes into a black hole and squirted out along the horizontal planes we see manifest. From the Galaxy, to the Solar system, the Earth, onto to the plant and the atom, it is the same story.

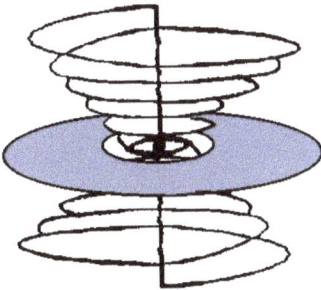

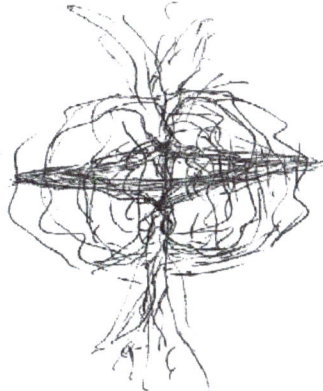

These `beings', due to their spinning motion create massive electro magnetic fields. In the case of Stars, their combustion creates electro magnetic and neuclear rays as well. It is this electro magnetism of creation which holds everything in its present form and acts as one of the main conveyers of formative force and activity between all the different layers. The planets play a special role in this. They have very predictable paths and cycles of relationship with each other. Therefore, as they move and alter the electro magnetic harmonics of our Solar System, at predictable moments, so manifestation alters accordingly. Hence the weather, plant sap flow and Human moods all move in a predictable dance orchestrated by the planets. As scary as this seems, my twenty years of observing these cycles has shown me beyond doubt that this is so.

THE PHYSICAL AND MACRO CARRIERS OF THE BODIES

One of the greatest gifts Steiner has given us is the associations he made between the elements of protein and the energetic bodies they carry into physical matter. This is one major meeting point between spirit and matter. The four elements of protein, Hydrogen, Nitrogen, Oxygen and Carbon are the basis of all organic chemistry and hence physical life.

Carbon, the basis of structure, is the seat of the physical form, Oxygen, the basis of life carries the Etheric body, while Nitrogen carries the Astral body providing sensation and Hydrogen carries the Ego which provides consciousness. Sulphur is a lubricator that helps these four basic elements combine into more complex substances (18, lect. 2). Seen from this perspective, chemistry takes on a whole new light. What follows is that only where these elements exist is it possible for these bodies to go.

Therefore the chemical characteristics of any 'environment', suggest the potential energetic activity of that environment. Steiner states "The composition of the external atmosphere is of such a nature as to furnish the ratio for the connection between the astral and etheric bodies and concurrently between their partners, the physical body and the ego" (6, pg. 157)

The internalised bodies can be characterised as follows:

As androgyny is still a rare occurrence amongst humans, so the basic level of polarity, level 2, remains an externalised function and we need male and female to come together to produce primary creation.

The Physical body is easily identified and anything that has physical substance has a physical body. It is localised to the planet Earth, and as such, our physical environment would be its seat. We are all Carbon based life forms. Steiner also pointed out how our physical bodies function according to a threefold system. Our head centres the activities of the Nerve Sense system, our chest centres the Rhythmic system and our abdomen centres our metabolic system. This threefold process links the modes of Astrology and level three of the Astrological model to the physical body and the physical earth.

The Etheric Body

The Etheric body, along with the other bodies, is not so easily seen. However the activities of these bodies leaves their mark or expression visually. The Etheric body is the 'energetic' body that imbues life and the capacity to grow and develop into all living forms. It comes from the Earth and has a potentially unlimited capacity to create growth. This is one aspect of the CHI body of Eastern philosophies.

It drives or energises the movement of fluids in general, which leads to the formation of mass in physical forms and supports the lymphatic and immune system of physical organisms. When the Etheric body leaves a living entity, death occurs. It provides a physical sensation of being up or down. When your Etheric body is strong you are full of life. Often a white glow spreading up to 4 inches from the physical body will be seen when an individuals Etheric body is strong. When it is weak you feel physically down and heavy. The individual with a weak Etheric body looks grey and feels drained.

The Etherics raw activity can be seen strongly in the first stages of growth where there are unshaped forms. For example in the first few months of a baby's life or the early stages of plant development. The early leaves of many plants are full and round and repetitive in comparison to the more adult leaves that are often indented and pointed. It produces watery shapes and curved forms.

The mineral kingdom cannot undergo this change through a death process as the Etheric body stays exterior. It remains as atmospheric forces of light heat and water. The Etheric body is only embodied by plant, animal and human. Crystals may appear to be an exception, however, they grow from additions to the outside of their form.

The Etheric body is carried by Oxygen. Around our planet oxygen is found in large quantities close to the surface and in diminishing quantities through up to the Ionosphere. At the cross over between the Stratosphere and the Ionosphere we meet it again as Ozone (O3). The Ozone layer shields us from many harmful cosmic rays most notably Ultra Violet and Gamma rays. These are mostly reflected off this shield enabling life to continue on Earth. While Oxygen is found in space and on other planets, it is usually chemically locked up and not free in the atmosphere. While I have not found many supporting quotes from Steiner, I have come to conclude that the Earth's Etheric sphere extends out to the edge of the Ozone layer. Hence this sheath being our life sheath, and as it disappears so life on Earth is threatened.

Steiner commented that the Etheric extended to the edge of the Blue sky which I understand to be approximately to the Ionosphere sheath. It is also interesting to note that Light and Warmth are developed from the Suns energy moving through our atmosphere. Karl Konig (7) suggests the four electro-magnetic belts within the Ionosphere are the seat of the Ethers. These belts are placed at 45kms, 100kms, 200kms and 300kms above the Earth. He was tentative, in his suggestions however it does fit with Rudolf Steiner's comments. This leads to an interesting picture.

As stated earlier the Atmosphere is created from the life processes of this planet and is the seat of the Etheric. As such, it is associated with the Elements sphere of Astrology and hence fourfold processes.

Within the Etheric realm four 'formative forces' or ethers have been identified. These are named Warmth, Light , Chemical and Life ethers. Each of these ethers has a intimate and polaric relationship with the physical elements of Fire, Air, Water and Earth respectively, to lift physical matter into specific living forms. (5)

ETHER	ELEMENT
Warmth	Fire
Light	Air
Chemical	Water
Life	Earth

The Etheric sphere resides within our atmosphere. The Ethers find their manifesting source in the 4 belts of the Ionosphere. Below this we have the Ozone layer and below this is the Stratosphere which is where the Elements reside. This develops the picture of the Ozone layer, acting as a membrane between these two inner spheres of the Etheric body

Ionosphere - Ethers

_____ Ozone

Stratosphere- Elements

The ethers are 'based' in the Ionosphere while the elements of Fire Air Water and Earth are 'based' from the Stratosphere down. The Ozone layer acts as a membrane between the two.

The Astral Body

The Astral Body imbues the physical and life form with sensation and movement. Commonly called the sense body, the Astral enables the living forms to experience their environment through taste, smell, touch and sound. From this sensation, the entity can then determine whether it feels good to be at that point or not. There is an 'animal' intelligence (instinct) carried by the Astral, which responds to stimuli - Does it feel good or not? The Astral body is also necessary for waking and sleeping to take place. Only animal and human kingdoms can experience awakening and sleep. In this sense the plant kingdom is in a constant state of 'human' sleep. Plant processes just sped up and slowed down according to the external stimuli of hot\cold, light\dark, wet\dry.

The Astral is associated with the element of Air and light. Its hallmarks in the animal kingdom are the degree of sensitivity and movement the animal possesses as well as the ability to internalise the breathing; the formation of true organs and their reactions to light.

In the plant world the Astral's influence is still present, however it works onto the plants from outside. The degree of leaf segregation, flowering and the development of poisonous nitrogen alkaloids and proteins, leads to hints of the strength of the Astral's activity. The Solanace and Legume families are two which draw the Astral activity closely into themselves. Hence their poisonous hallucinogenic effects. The states of consciousness some substances from these families produce in Humans are experiences of the Astral realm.

In Humans, the Astral influences the formation of the organs and the processes as in animals, however it also makes its presence felt through our psychological sphere. Emotions, dreams, imaginations and 'psychological ' swings are all related to the Astral body's activity.

The Astral body resides cosmically in the spheres of the planets. As such, it has a sevenfold character and is influenced strongly by the movement of the planets. The seven energy centres of the body referred to as the Chakras are 'organs ' of the Astral body.

Spirit / Ego

The Ego (the spirit) imbues the sensitive life form with intelligence and individuality. As separate from the dream consciousness of the astral intelligence -instinct, Ego intelligence is related to processes of thought and deduction, as well as remembering and forgetting, processes that enable its recipient to consciously determine action and response. Once the response has been determined, the Ego imparts the degree of commitment by which the action is carried out. Through this action, body heat is created indicating the Ego's relationship to the element of Fire and the Warmth ether.

The Humans are the only kingdom presently with the potential to internalise this body. This bestows upon us the potential for self consciousness and to make free choice with regards to the sensations and instincts the Astral body experiences. The degree this free choice is actually exercised may indicate the individual's stage of 'energetic' awareness.

In the other kingdoms of nature the Ego still works from outside, as a collective function or group soul. In animals, the group soul is evident in the flock of sheep or birds, the school of fish and the pack of dogs. In both animals and plants the species firmly connects the individual plant to the collective Ego which resides as a Cosmic level in the Fixed Stars.

With Humans internalising the Ego, our connection through race and blood ties are allowed to begin to breakdown. We become energetic individuals able to form associations through ideology, faith and individual preference, irrespective of race and blood background.

These are the four major aspects which influence life's functions. In creation there is very little black and white and there are many shades of grey. In observing how various entities work with these four bodies , there are often occasions when an entity is in a grey area. Certain animals may seem very plant like, or almost human, while even humans can sometimes take on a plant like quality.

We are therefore presented with the following picture :

Body	Macrocosm	Steiner	Astrology	Element
World Ego	Galaxy	12 fold	Fixed Stars	H
World Astral	Solar System	7 fold	Planets	N
World Etheric	Atmosphere	4 fold	Elements	O
World Physical	Life on Earth	3 fold	Modes	C

From here it can be surmised that that Ego and Physical body have a specific creative relationship and that the Astral and Etheric create a formative harmonic as well. As we will see and from Steiners writings we note this is so. While these are the cosmic seats of the energetic bodies, the four kingdoms each personalise the bodies in their own way.

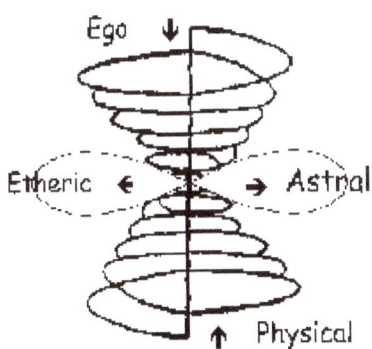

Soul World
Stage 2

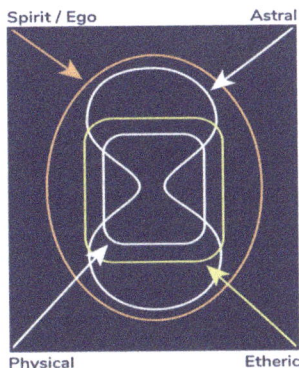

Manifest World
Stage 3

THE ENERGETIC BODIES & THE KINGDOMS

Mineral Kingdom

The Mineral kingdom has a physical form with the Etheric, Astral and Ego remaining external to the physical form. This means these bodies work directly from the World sphere mediated through the elements of warmth, light and moisture.

Plant Kingdom

The Plant Kingdom has a physical form and an internalised Etheric body, and so it can grow and expand as well as die. However the Astral and Ego remain external in the World spheres. So, while plants will expand and grow, it is usually external influences of light, moisture and heat that determine their final form. Flowering and seeding are processes bought about by the astral and ego acting onto the plant externally, through the Light and Warmth ethers of the Etheric body.

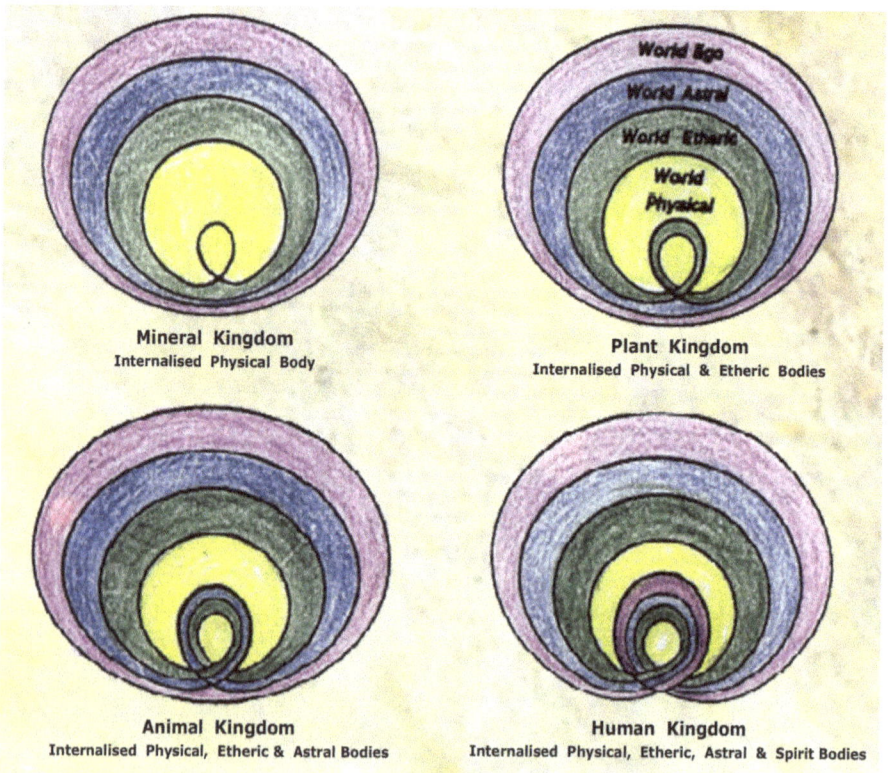

Mineral Kingdom
Internalised Physical Body

Plant Kingdom
Internalised Physical & Etheric Bodies

World Ego
World Astral
World Etheric
World Physical

Animal Kingdom
Internalised Physical, Etheric & Astral Bodies

Human Kingdom
Internalised Physical, Etheric, Astral & Spirit Bodies

Animal Kingdom

The Animal Kingdom embodies the Etheric and Astral, inside their physical bodies, while the Ego stays external. This allows them movement and sensation, but not a high degree of 'consciousness'. So, if the food or temperature is not right in one position, then they will move to another. While a degree of freedom is expressed here, it is usually based on sense experience or habit and instinct. The degree the astral is internalised is indicated by the formation of the four organic systems of lung, kidney, bladder and heart.

Human Kingdom

Humans have the potential to embody all four spheres. The Physical , Etheric, Astral and Ego. In so doing, we become potentially conscious and self determining individuals. The Ego has been the last faculty to fully incarnate and the history of, especially the last 5000 years points to its increasing influence in the life of the human. A strong awareness of matter took place in Persia and Egypt. The development of the Mind though Greece, and the subsequent development of science, are all symbols of mans growing awakening to the Ego. A trail to the awakening of the individual, as separate from the tribe or race, parallels this development.

External Bodies

If the bodies are externalised it does not mean they cease to function. They function directly from Fixed Stars (Ego) , the Planets (Astral) or through the Ethers or Elements (Etheric) from the outside onto the individual form.

In plants the Etheric body, which has an outward movement, is contained and balanced by the inward moving Astral and Ego, externally. In the case of animals, the Astral (still inward moving) enters the body and can balance the Etheric from within. Nature spirits and Devas are really the collective Astral/ Ego of the particular plant species. The individual plants are energetically linked to this common Astrality and Ego.I t was Wilhelm Pelikan who said manifestation: IS in the mineral kingdom, LIVES in the plant world, EXPERIENCES through the animals, and COMPREHENDS ITSELF through humans (10 Vol 1).

Putting all the above information together, a picture of life's interrelationships as a Microcosm of the Macrocosm, develops as follows. This diagram outlines the internalisation of the World energetic bodies into the personalised kingdoms of nature through their various associations. Ideally all four bodies work harmoniously according to the makeup of any particular plant, animal or human form. In this state, optimum health will be reached and maintained.

From this point of view pests and disease in plants, and illness in people can be presented in a new light. Biodynamics, like Steiner's medicine, holds the view that if there is any imbalance between the bodies, disease or pests of some form will manifest.

The Double Spiral

Just as the Astrological Theorem can be seen in the Macrocosmic environment, when we come to manifest life, the same patterns are present. Some examples of this have already been presented. The double spiral has the top half being the Macrocosm and the bottom spiral being the microcosm.

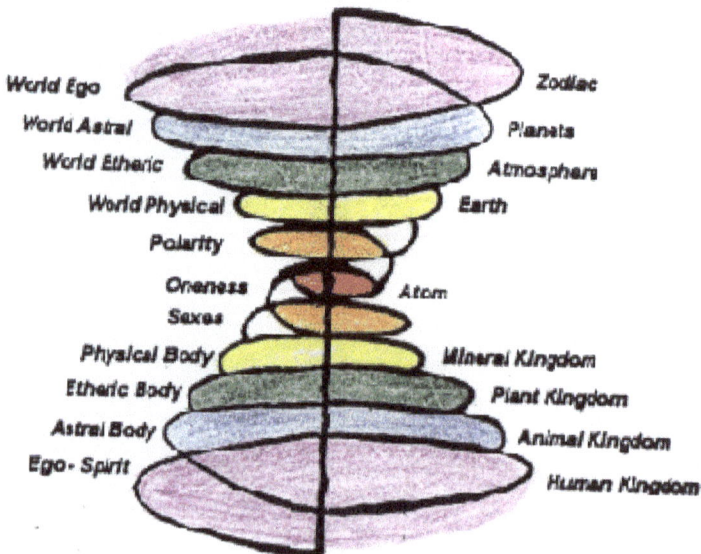

World Ego — Zodiac
World Astral — Planets
World Etheric — Atmosphere
World Physical — Earth
Polarity
Oneness — Atom
Sexes
Physical Body — Mineral Kingdom
Etheric Body — Plant Kingdom
Astral Body — Animal Kingdom
Ego- Spirit — Human Kingdom

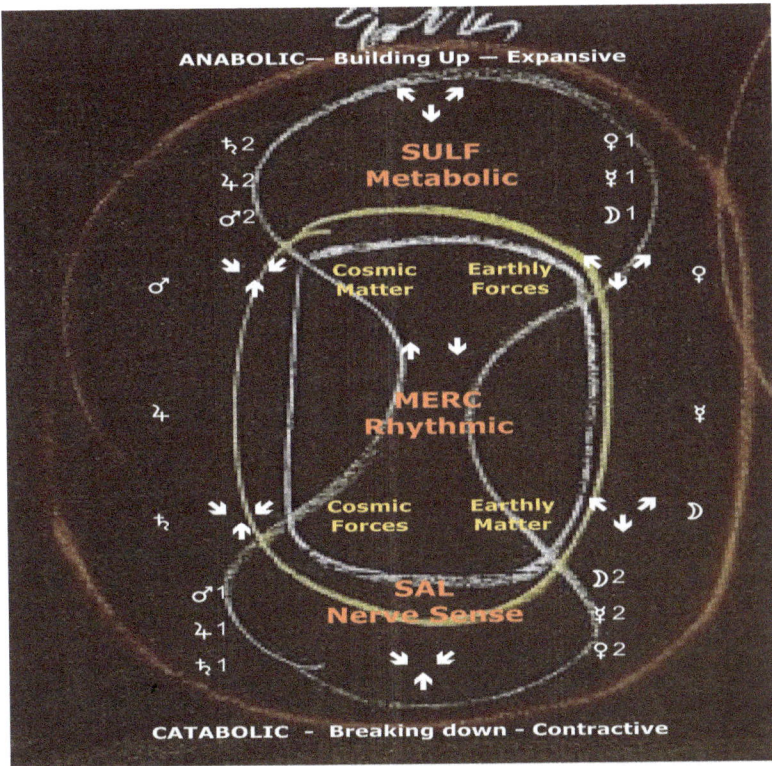

ANABOLIC— Building Up — Expansive

SULF
Metabolic

♄2
♃2
♂2

♀1
☿1
☽1

♂

Cosmic Matter   Earthly Forces

♀

MERC
Rhythmic

♃

☿

Cosmic Forces   Earthly Matter

♄

☽

♂1
♃1
♄1

SAL
Nerve Sense

☽2
☿2
♀2

CATABOLIC - Breaking down - Contractive

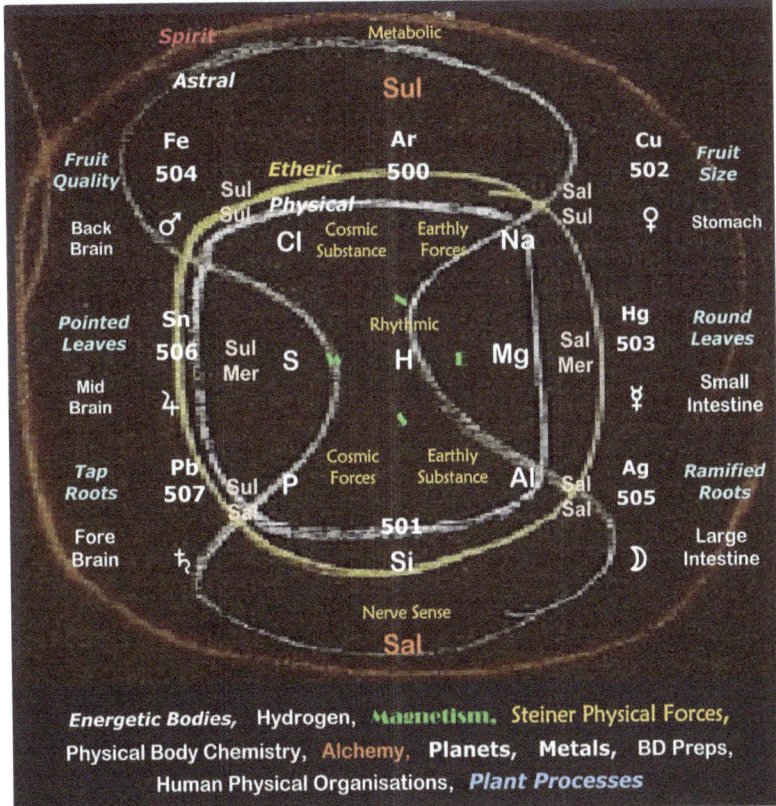

Spirit   Metabolic

Astral   Sul

Fruit Quality   Fe 504   Ar 500   Cu 502   Fruit Size

Etheric

Back Brain   ♂ Sul Physical   Sal Sul ♀   Stomach

Cl   Cosmic Substance   Earthly Forces   Na

Pointed Leaves   Sn 506   Rhythmic   Hg 503   Round Leaves

Mid Brain   ♃   Sul Mer   S   H   Mg   Sal Mer   ☿   Small Intestine

Tap Roots   Pb 507   Cosmic Forces   Earthly Substance   Al   Ag 505   Ramified Roots

Sul Sal   P   501   Si   Sal Sal

Fore Brain   ♄   Nerve Sense   ☽   Large Intestine

Sal

Energetic Bodies, Hydrogen, Magnetism, Steiner Physical Forces,
Physical Body Chemistry, Alchemy, Planets, Metals, BD Preps,
Human Physical Organisations, Plant Processes

THE CREATIVE FORMATIVE FORCES

The process of creation is the manifestation into the physical dimension of the Earth through the combined influences of the Macro spheres of our environment. The Fixed Stars are a cluster of stars and galaxies emitting immense concentrations of radiations. These radiation sources are relatively constant in both the intensity of their output and their place of output relative to us. The Fixed Star movement in relation to us is minimal. Hence they are called Fixed Stars. This consistent state of source energy sets up a constant field of force, at which the Earth is a 'centre'. It is inside this constant field of force that the movements of the planets in our Solar System alter the electro-nucleic-magnetic (ENM) radiations from the stars. The planets movements are best described as rhythmical. They continually circle the Sun at set timings and at set distances. Therefore many of the relationships between the planets reoccur at very predictable timings. It can be said therefore that the pulses or rhythms established by the movements of the planets establishes a constant movement of energy which, over large periods of time, could be considered constant. Therefore, we have a circumstance where an internal fluctuating yet predictable reality, is established within a constant field of force.

Due to its innate stability the Fixed Star realm or Zodiac sphere is considered to be the sphere of the archetypes. Stars emit EM forces constantly and so these forces can be considered the formative patterning's upon which life forms are developed. As an image, it has been suggested as early as the 1500s by Paracelsus that each plant species owes its archetypal form to an individual star.

In reality there are several influences making up any single entity. The main point here is to see the Fixed Stars as emitting the original formative impulse.

To recall, it should be mentioned that we live in a spherical reality. Earth is the subjective centre of our Universe and the 'layers' outlined

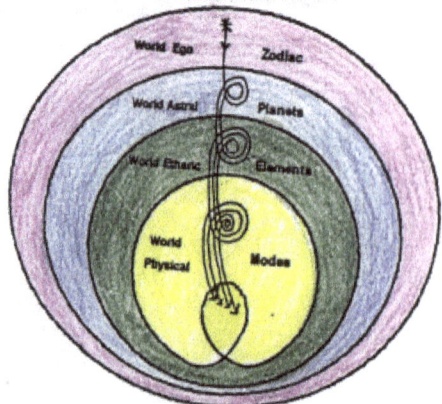

elsewhere in this publication are spheres around us, like rings of an onion.

As the formative forces move towards the Earth from the Fixed Stars, they pass through the Planetary sphere - the seat of the World Astral Body. The mixing of the Energetic formative force with those present in the planetary sphere allows the Star forces to 'pick up' the Astral formative forces. These then move through the Ionosphere and our Atmosphere - the seat of the World Etheric body- to 'pick up' the World Etheric formative forces, and onto the Earth to become entwined with the physical formative forces carried through the light, heat, moisture, and chemical elements.

When we talk of formative forces we therefore need to be clear as to what particular formative impulse we are dealing with. It would appear that traditional biodynamics has become possessed with the idea of Etheric formative forces, especially with plant growth and have lost the differentiation of the other activities. This is where Glenopathy differs quite markedly from the traditional approach. We work with the four energetic bodies as individuals and note that they are continually active within and from without the plant. When a energetic body incarnates into physical form it works directly for itself, however the indirect or 'World' influence of a body also works at the same time. Hence we have internal warmth coming from within our body and external warmth coming from the Sun. When a body is not incarnate, it works indirectly from the outside and, also through relationships it has, to parts of the incarnated bodies.

Plants are a good example. They have an internalised Etheric body, however the astral body is generally external. Only in poisonous plants does it enter a certain distance. While the Spirit is always external, however it works through the Cosmic Force process active at the Physical Formative Force level.

The fourfold division of the plant has been established through the work of both Maria Thun and Agnes Fyfe. They found that if plants were worked when the moon was in certain elemental constellation, the corresponding part of the plant was enhanced. The fire constellations enlivened the Seed and fruit. The air constellations encouraged the Flower. The water constellations encouraged the Leaf. The root was encouraged by working in the earth constellations.

Through association we see that Fire relates to the Spirit and the Seeding process. So in plants the Spirit impulses coming from the Fixed Stars find their way into the plant processes through the mediation of the Warmth ether of the Etheric body, along with the various 'Spirit' minerals like Silica Phosphorus, Chlorine and so on.

Likewise the Astral body works onto plants from the outside, directly from the planets, yet its influence works indirectly through the etheric body through the Astral's connection, to the etheric bodies' Light ether and the PFF Cosmic Substance. Minerals such as Silica sand, Arsenic, Cobalt and Chromium all stimulate the Astral activity.

This fourfold level runs through many different aspects of our life and environment.

There are a series of layers in creation and at each layer Dr Steiner uses a different name to help identify the different carrier of that activity.

As we move to the physical body we see that in lecture 8 of Agriculture, Dr Steiner talked clearly about Cosmic Forces, Cosmic Substance, Earthly Forces and Earthly Substance. He also outlined the workings of Calcium and Silica in two different ways working in the environment , which has caused massive confusion for following generations. I have offered an answer to this confusion in 'The Planets in the Agriculture Course'. Suffice to say for the moment, I see these two aspects of both Calcium and Silica as two parts of one cycle. So we have a Cosmic Silica and a Earthly Silica as a cycle and as well as a Cosmic Calcium and Earthly Calcium as another cycle, which interact with each other.

Later chapters of this book looks in more detail at the connections between these different levels. However, it should be mentioned here that the relationships can be outlined as ...

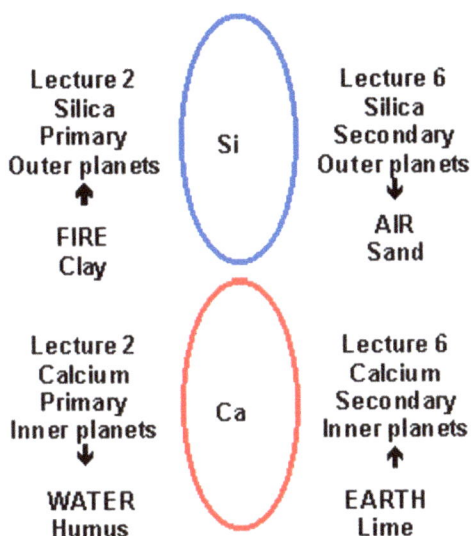

Lecture 2 Silica Primary Outer planets ↑ FIRE Clay	**Si**	Lecture 6 Silica Secondary Outer planets ↓ AIR Sand
Lecture 2 Calcium Primary Inner planets ↓ WATER Humus	**Ca**	Lecture 6 Calcium Secondary Inner planets ↑ EARTH Lime

'BEING'   **'MANIFESTATION'**

STARS

PLANETS

ETHERS	Warmth	Light	Chemical	Life
ELEMENTS	Fire	Air	Water	Earth
ENVIRONMENT	Cos. Si.	Ear. Si	Cos. Ca	Ear. Ca
BODIES	Cos Forces	Cos. Mat. Ear. For.		Ear Mat.

This fourfold activity within the layers of life can be taken further, from the physical bodies into the fruits, seeds, cells, bio-chemistry and DNA. The forces active in all these processes have originated in the wider Universe, our Galaxy and the Solar System. They work down and use the 'players' of the various layers of creation.

Each of these streams of force can be influenced by the substance of any particular layer. See pg. 105. RS commented that Saturn worked more strongly when the atmosphere was hot, while the forces of the Moon worked more strongly in a wet environment. This indicates that the working of any particular force is dependant on the quality and quantity of the carrier elements of any lower spheres.

ARCHETYPAL & SUBJECTIVE REALITIES

An important distinction to make is the relationship between the 'real' subjective experience of reality and how this corresponds to an Archetypal 'reality'.

A subjective reality is what we directly experience based upon some reality, be it stimulated by the planetary positions right down through to the weather. Our subjective experience in itself may only be a partial 'true', however, it is nevertheless meaningful and in need of serious consideration. With Astronomy for example, we Humans while living on Earth, experience creation as if we are at its centre, which of course we are not. Nevertheless the Sun appears to revolve around us, not the other way round. The experiences we have of the Sun being secondary to the Earth, such as sunrise and sunset, form part of our subjective reality. This is what we experience or perceive from our point of view. The Astronomical truth is that we revolve around the Sun, yet our subjective experience of the Sun revolving around us has a dominating position in our experience of reality.

Having said this, a close investigation of the planetary influences, from a heliocentric perspective, also reveals that they have an effect upon the way life manifests on Earth. This occurs simultaneously to our Earth centred (geocentric) experience. The point is that both the geocentric and heliocentric experience of the planets form part of our subjective experience of a reality. As will be shown later, we can however delineate a difference in the 'quality' of these influences.

An archetypal reality is a principle, rule or 'law', which can be found in some aspects of life. This law, or principle, is based upon some physical or cosmic reality. However it may not have any direct relationship to the subject at hand. In the context of this work the archetypal realities we meet are often an Astronomical phenomena whose principles can be found in some element of life we wish to explore . The rules of the original phenomena are moved and applied to some other phenomena. E.g. the seven year cycles of human growth so commonly accepted, are actually based upon a fourfold division of Saturn's 28 year cycle, or when RS associates the human senses to the twelvefold nature of the constellations, or when we have the 12 signs of the zodiac developed from the astronomical reality of the 12 constellations. These principles are meaningful structures used to organise our experiences of life, however in themselves they do not relate to a direct reality but rather a fundamental patterning or archetype taken from somewhere else.

The archetypal principles and subjective reality often interact with one another and can be seen in the seven year cycles mentioned above. Saturn has a basic cycle length of 28 years. It has been found that the 4 quarter phases of this cycle indicate significant moments in the development of an individuals personal responsibility. (See "Parenting as a Saturnine Pleasure"). Therefore we have come to develop a general or archetypal theory of growth based upon the seven year cycle. In reality though it has been found that the trigger and timing of these events relate to when Saturn has its exact 90 or 180 degree relationship back to its birth position. This relationship does not occur every seven years exactly. Saturn has its actual relationship back to its birth position anywhere between every 7 & 9 years. So the archetypal law shows us an overall pattern, which provides a general human effect, yet the subjective reality of the individual gives us the actual event for them. Both points of view have some relevance and can be considered correct.

Another example, which we will look at in greater detail later, is the relationship of the constellations of the zodiac to the signs of the zodiac. Essentially the reality is that there are twelve constellations of stars standing behind the ecliptic of the Sun. These are clusters of energy generating stars which have moved very little in a long time. This twelvefold archetypal nature of the constellations (which are a galaxy phenomena) has been 'picked up' and then placed upon the path of the Sun around the Earth, (which is a Earth centred Solar system phenomena). The starting point of this 'Earth' zodiac is the northern hemisphere spring equinox. The 12 divisions created are then called the signs of the zodiac. They are a manifestation of an archetypal pattern of creation being taken and applied to another circumstance. The function and effect of this is another matter and will be addressed later. The point here is that this 12 fold archetypal law of life, drawn from the constellations, is able to be moved around and applied to almost anything. We have seen how the 12 fold law relates to the Energetic level and thus applying the 12 fold law stamps a energetic dimension onto the subject. Applying the 7 fold law stamps and Astral focus upon the subject and so on.

So an archetype, in this sense, is a basic law of creation. These laws, whatever they are, hold good for the new realities upon which they are built and some relevance will arise from them. The nature of the law indicates at what level or arena the subsequent information will relate too. Any process of life can be divided into twelve ,seven, four, three or two fold processes as RS continually shows. The Astrological model therefore provides us with a Universal theorem which outlines many of the basic archetypal laws of the galaxy, as the galaxy presents them to us. The patterning allows us to determine where these laws work and what their influence is when they are applied to subjective life experiences.

Interestingly, Steiner more often than not uses the archetypal division of things rather than the real. Take for example his twelve fold division in the 12 senses work. It can be suggested this activity is describing an Egoic level of the senses.

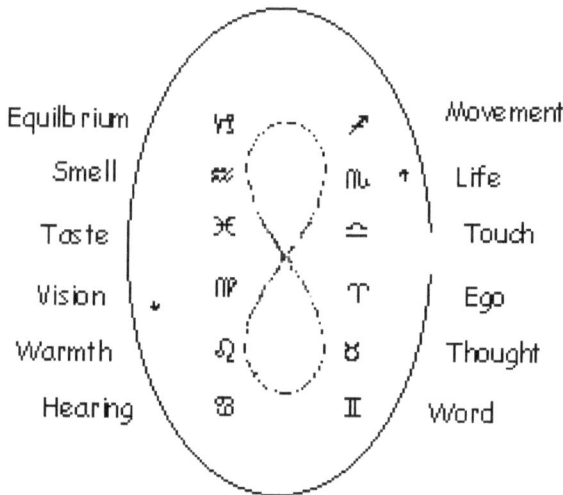

The Twelve Senses

When RS uses the planets in the biography work he is talking on an Astral level. Hence he is discussing the soul or psychological development of the individual. B Lievegoed is his book "Phases" also uses the archetypal planetary patterns rather than the real ones. This book gives the mid-life crisis as taking place at 42 based on the archetypal rhythms of the planets. In reality though the mid life crisis period runs at present from 37 - 45 based on the real planetary movements. In this regard a wonderful book called 'The Cycles of Becoming " by Alexander Reperti was written for Astrological circles in 1978. This book outlines in detail the exact cycles and rhythms of the planets with their influences on human development. This is the Astrological equivalent of Lievegoed's book. Well worth the study.

Anthroposophical critics of Astrology can scarcely 'complain' about the use of archetypal patterning for the development of the signs of the zodiac when their whole reference system is built upon the exact same process. The signs in this sense are a brother of many of their understandings.

Archetypal patterns have great relevance. They show up everywhere in Astrological study. The Astrological theorem shows the complexity of interrelationships between all the various parts of creation and indicates the laws for the manifestation of almost all of creation. In the field of human psychological development, and other subjective realities, one must not overlook the experience of the real planets as the activators of development. The planetary positions of the natal or progressed charts give the exact timing of the events of life. The archetype can give us their meaning.

Both the archetypal and the real phenomena can be used and are both "right".

Planets		Zodiacal Constellations			
		+		−	
♄	Saturn	♒ Aquarius		Capricorn	♑
♃	Jupiter	♐ Sagittarius		Pisces	♓
♂	Mars	♈ Aries		Scorpio	♏
☉	Sun				
♀	Venus	♎ Libra		Taurus	♉
☿	Mercury	♊ Gemini		Virgo	♍
☽	Moon	♌ Leo		Cancer	♋

Stage 1 – Zodiac

THE BIODYNMAIC VORTEX

LEVEL ONE - Individual spirit or the Earth

The first level is the point in between the macrocosmic and the microscopic universes. (See 'Oneness' on the Double Spiral) It represents both the energetic unmanifest state of oneness and the Earth as a being in its own right. Life only begins to manifest when the polarity level is reached. This can be experienced as a state of consciousness and can be seen as the individuals inner contact with the infinite. This is the spirit spark which holds the ultimate impulse to come into creation. As we hold our consciousness with this part of ourselves, we again feel our contact with the outer collective spirit outlined by level 6 the Zodiac.

LEVEL TWO - Duality - Calcium & Silica - 2 fold

In her adventures through Wonderland, Alice experienced the large and the small. In Bio-Dynamics, the understanding of opposites is also a major cornerstone of the philosophy. Balancing the opposites is necessary to create harmony. These opposites are referred to in many esoteric systems as "female" and "male"- Yin and Yang in the Taoist tradition. The female aspect represents the receptive nurturing, fertility principle, while the male is assertive and more outwardly active leading to the nutrition forming processes. Philosophically, these are seen as the moon force and the sun force . Rudolf Steiner points out these primary streams of force are mediated in living beings by the substances which focus their activity on earth. The female stream of the Earth is mediated by the Limestone of our planet. While the male cosmic stream is mediated by the Silica in the Earth.

These substances act as the carriers and anchors of these 'Macro' streams. It is along these major currents which all else flows. All of the influences identified in the other layers of the BD vortex use these two streams as their pathways. It is almost best to picture them as two separate trains running on over the same countryside but on different networks. Later on we will see how the other layers work in with these. Early in 'Agriculture' RS outlines how the planetary formative forces from the outer planets, Saturn, Jupiter and Mars impact the Earth through the Silica elements, while the inner planetary forces, Venus, Mercury and Moon are carried by the Calcium substances radiations.

Plant growth

In the first two lectures of Agriculture, Rudolf Steiner outlines the complex manner in which these two primary forces function, to create plant and animal life.

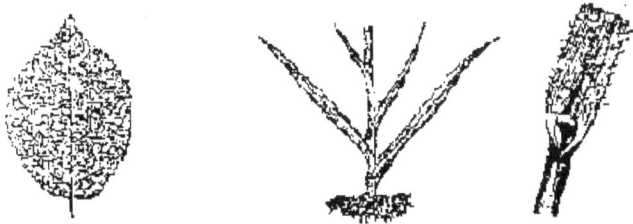

These same influences show up in plants tendency towards tap roots or netted root systems

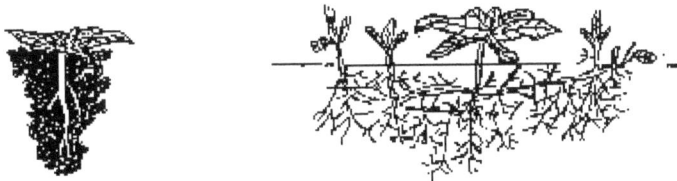

These two basic force processes dominate plants, and also show up in the dual tendency to reproduce or to focus on the ability to provide nutrition for higher life forms. The forces which provide the power of reproduction and growth are carried by the Calcium stream and the "inner" planets, while the Silica and "outer" planets mediate forces which produce plants with a high nutritive value. "Silicon opens up the being of the plant to the expanses of the Universe, it awakens the plants senses." (Agriculture)

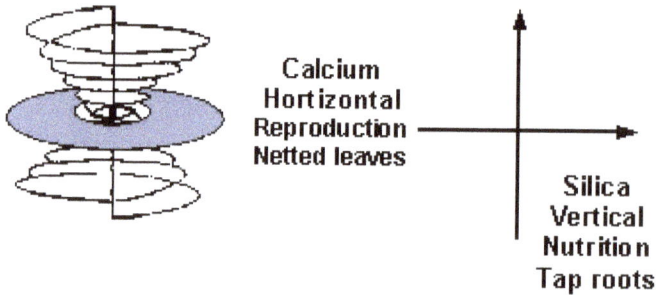

Calcium
Hortizontal
Reproduction
Netted leaves

Silica
Vertical
Nutrition
Tap roots

We saw earlier how the cosmic creative process eventually manifests a vertical and horizontal plane. Plants follow the same formative structures. While these can be seen in clear individual examples they also cross over. The grasses which have vertical veining right through their leaves also have a high silica content in them, pushing the Silica forces into an area which in other plants is dominated by Calcium forces. You could say the silica stem process dominate in these plants.

In lect. 1 & 2 of Agriculture, we are told that for a plant to grow, the Silica stream carrying the forces from Saturn Jupiter and Mars, is first attracted to the soil through the Silica (sand) in the earth. These are the forces carrying heredity and the will for the furtherance of the species. The mid winter process anchors these silica forces to the earth through a process RS calls 'crystallisation'. After mid winter especially, these forces begin to move upward from the soil. They combine with the seed to initiate the plant on its journey of growth. The clay in the soil acts as a kind of mediator and allows this force to move upwards. If the silica content of a soil is too great, this force will remain in the soil, encouraging the tap root of the plants. The clay enables this force to move upwards forming strong stem growth on towards the ultimate reseeding of the plant.

The calcium stream is drawn from above and anchored to the soil through the lime and humus in the soil.

At germination the calcium stream combines with this vertical silica push to create the physical quality of the cell division and tissue formation and shows also in the rhythmical spiralling development of leaf formation. This process is aided by the amount of humus in the soil which leads to the mass formation and size of the leaves and the

92

abundance of flowers. This highlights the roll of compost in the Biodynamic venture. Without compost, the inner planetary forces of reproduction and mass formation can not be drawn close enough to the Earth to give the Silica stream enough substance to reach its ultimate goal of seed formation.

The biennial plant gives us a picture of this process. In the first year the tap root is formed. The upward thrust is contained in the root. The Calcium leaf formation continues to spiral, generally with little leaf differentiation or stem development. The silica force is not yet moving upwards, but is being contained in the soil while the rosette is formed. In the second year, the silica force begins its move to seed. It bolts upwards and with it the leaves are drawn up along the stem into space. Greater leaf differentiation is evident. As it moves into space and in contact with the light and warmth, pushing into the World Astral sphere, the leaves become smaller and flowering occurs. Fruit set follows and the plants dies as it contracts all its forces into the new seed.

When the calcium stream and humus are not available, the seeding process will begin too early, leading to premature ripening. Plants can become spindly and creeper like. Observe the plants growing in deserts or sand dunes. Similarly when Silica is lacking, plant roots become more ramified, stems become thicker with flowering and fruit set being minimal. Observe the growth on swamp land or lush dairy pastures.

This is an unusual concept at first, however on further investigation, it does prove to be a most fundamental one. It is the task of the Biodynamic practitioner to balance the interrelationship between Calcium and Silica throughout the year to produce our desired results.

The Great Biodynamic Contradiction

There is some disagreement in Biodynamic circles about the interpretation of Steiner's agricultural lectures. This is due to RS appearing to contradict himself in what he says about Ca and Si in the second lecture and what is said in Lecture 6. He outlines two different directions for the Calcium and Silica streams. The answer may lie in the awareness that in lectures 1 through 3 he is describing cosmic formative

o   r
force

Lecture 2		Lecture 6

Lecture 2
Silica
Primary
Outer planets
↑

Si

Lecture 6
Silica
Secondary
Outer planets
↓

FIRE
Clay

AIR
Sand

Lecture 2
Calcium
Primary
Inner planets
↓

Ca

Lecture 6
Calcium
Secondary
Inner planets
↑

WATER
Humus

EARTH
Lime

'BEING'   **'MANIFESTATION'**

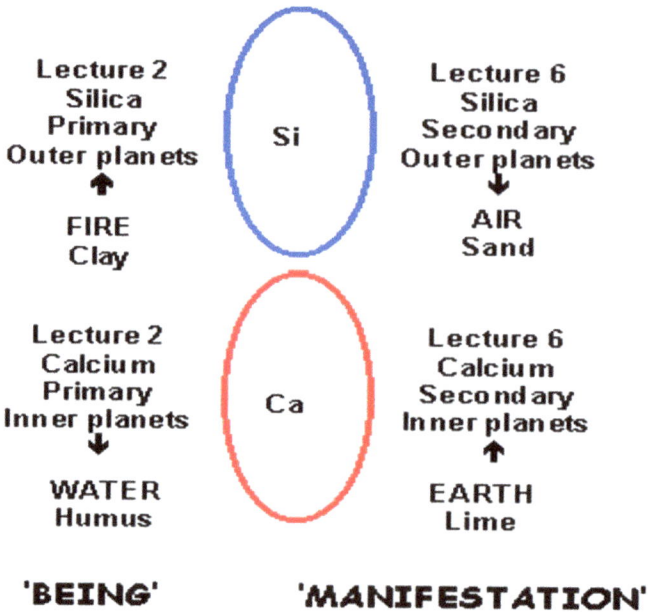

processes of nature, while in Lecture 6 he is dealing more specifically with physical problems such as fungal attack.

When the Ca and Si processes are seen as circular processes, with both an incoming and outward moving phase of their activity, this contradiction disappears. Preparations 500 and 501

With the Calcium and Silica forces being primary activities in nature and essentially the basis of any balance, Dr Steiner outlines two preparations that help to moderate any imbalances in their formative activity in life forms.

Preparations:.... " One to make you Larger... and One to make you small..." so the caterpillar says to Alice.

These specially mixed preparations, in combination with any physical adjustments, such as tree planting, composting and crop rotations, help to establish the inherent harmony of any environment. The balance of the growth processes, characterised by Sun and the Moon, is achieved through the use of a preparation of cow manure (female, Moon) "500" and a preparation of Quartz crystal (male, Sun) "501".

Preparation 500 (Horn Manure)

This preparation is made by placing fresh cow manure inside the horn of a cow. This is done in early Autumn. The horn is then buried underground into the living soil and dug out in spring before the soil begins to dry out. Place the horn so that water does not collect in the horn while in the ground. Place the horns in good soil not in the subsoil. Make sure there is not too much organic matter in or around this pit as worms will be attracted and they LOVE this mixture. Place some manure on the top of the soil covering the horns to attract any worms into it that rather than into the horns lower down.

Its effects

With this preparation we are bringing Metabolic processes into the Nerve Sense system / Soil. Physically, the cow manure preparation activates the soil micro flora and their digestive processes. These silent chemists liberate minerals from the soil to enhance root and general plant growth. Energeticly, the etheric or life giving forces of the Earth and plants are enhanced, thus allowing them to grow more lushly. This leaves the impression of a stronger vigour about the entire landscape. Used to excess, however, symptoms of over fertilisation develop in plants. Leaves become watery and begin to droop under the weight of their own mass. As 500 processes use humus as their 'plate' of activity, excess use can deplete the soil of organic matter through over activating the soil biological processes. It is therefore important to replenish organic matter at the same time as using 500. The more organic matter and humus in the soil the better 500 works.

An excess however, would seem a physically impossible feat, as a two inch diameter sphere of this preparation - physically just cow manure - is applied to 1 acre of land. Symptoms of excess however, can be observed in some cases after three or four applications.

Preparation 501 (Horn Silica)

This preparation is made in essentially the same manner, however it is made from ground up quartz crystals. The crystal is ground to a talcum powder consistency before use. Slightly moisten the powder before putting it in the horn and then bury as before. This time the horns are placed into the ground in spring and left in the soil until mid autumn.

Using it - **Here we are brining Nerve Sense activity into the Metabolism** The quartz preparation is applied to the leaf and flowers of the plant and intensifies the quality of light within the general environment. This strengthens the structure of the plant and enhances the nutritive value. Flowering and fruiting are enhanced as well as a general decline in the susceptibility of plants to fungal attack. One of the few known uses the plant has for silica is to strengthen the cell wall. This may account for a portion of fungal prevention. However the infusion of an environment with a fine spray of quartz crystals, intensifies the plant's use of available light, thus effectively changing the whole growing environment. The damp wet conditions, a fungus likes so much, no longer exists.

One gram of pulverised quartz preparation is used to cover an acre of land with a fine spray. General effects to be noticed are that plants stand more upright and are able to respond to dry/ wet changes more readily. Flowers are of a deep colour and scent and the fruit will ripen more sweetly and earlier. There is greater resistance to fungal problems. Excessive use can be noticed after two successive applications, in close time proximity, causing sunburn of crops and desert like conditions. Fruit size can also be reduced. It is best to spray this preparation generally before 11am in the morning. More recently afternoon spraying has been suggested if the contractive aspects of 501 wish to be enhanced.

This preparation should only be used after the Horn Manure preparation has been applied at least twice.

These two preparations are used in conjunction with each other. Used individually they can create disorder but together they go along way to recreating an environment resembling an experience of the wild forest in your own backyard.

Application.

Once the preparations have been in the soil there is another step suggested by Rudolf Steiner. The prescribed amount of preparation (25gms) is placed in a bucket with 3 gallons of hand warm water

(Rain water if possible). This mixture is then stirred in alternating directions, creating a deep vortex in each direction before changing to go the opposite way. This stirring is carried out for one hour. The "horn manure" is then applied to the soil in large drops while the "horn silica "is applied to the leaves of the plant in a fine spray.

This stirring process activates and releases the forces contained in the basic preparations. This process essentially homeopathically potentises the preparations to the first potency.

The amount of substance or bacteria present in this process can not account for the effects produced. It is the forces contained within the 500 and 501 which are released into the water and then applied to the soil and plants. The preparation making process concentrates the 'energetic ' forces already present in the primary substance and present in the environment during the 'composting' period. So during stirring and spreading, it is not the matter, but the forces behind the matter which works for the Biodynamic grower. Science is now telling us everything is energy. Biodynamic growers work with and focus the abundance of unseen energy in life to create environments suitable for plant growth. The basis of health is balance and activity. These two preparations provide the primary basis of both of these.

BDMax Sprays

BdMax have developed and activated these preparations further homeopathically. Etherics 1000 and SilicaMax, do not need further stirring in the way outlined above. They are diluted and a simple mixing stir is enough. They have the added advantage of maintaining their activity for a number of years in storage. The original preparations have special storage requirements and need to be spread immediately after stirring.

Commercial results since 1989 suggest these BDMax compounds act more vigorously than the original preparations. see (Case Studies on the website)

Duality

Wherever there is duality identified in nature it is related to level 2 of the spiral. Similarly when using these preparations other manifestations of either pole can be used to enhance the overall effect of the preparation.

501	500
Male	Female
Summer	Winter
Morning	Afternoon
Heat	Cold
Light	Dark
Dry	Moist
Expansion	Contraction
Silica	Calcium
Clay	Humus
Sun	Moon
Waxing Moon	Waning Moon
Ascending Moon	Descending Moon

Level 3

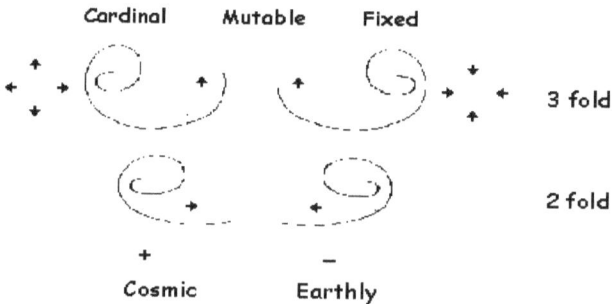

In Astrology this level is called the Modes and in modern science is described as Thesis, Antithesis, Synthesis. Generally in Astrology, the Cardinal pole is seen as the expansive outward moving pole and therefore male in nature, while the Fixed pole is seen as contractive and therefore female in nature. In Biodynamics however these two poles are reversed. The Astrological interpretation is a psychological view while the Biodynamic view is of physical processes and physical bodies. This polarity of activity can be seen later in the Astrological Model when we look at the inner workings of the 12 fold zodiac.

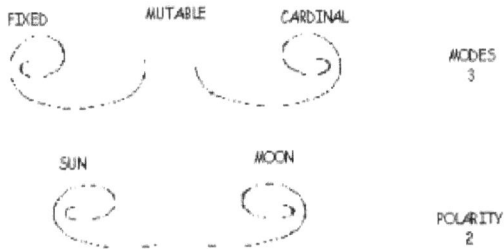

R.S. points out that physical forms, especially the more developed animals and human forms, are divided into three specific regions. The head region is the centre of the Nervous system. The Diaphragm is the seat of the Rhythmic system containing the lungs and the heart / circulation systems. The torso is the centre of the Metabolic system, incorporating the digestive and reproductive systems.

The nerve sense works in from the head down. The Metabolic forces work upwards from the feet as it were, while the Rhythmic forces are developed and harmonised out of them both. The more developed the life form the more individualised these systems become.

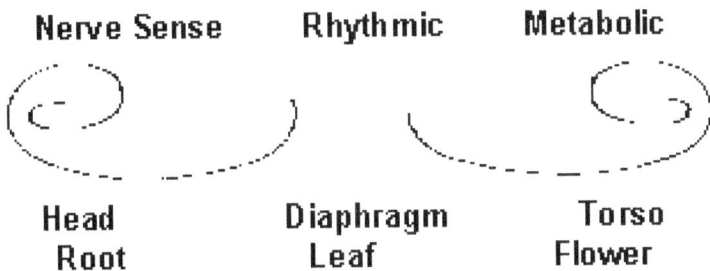

When describing a farm and plant growth, RS refers to the "agricultural individuality' as if it where a human standing on its head. The Nerve sense region of the plant is the root and below the Earth. The Earth's surface is the Diaphragm " while we and all the animals live in its belly".

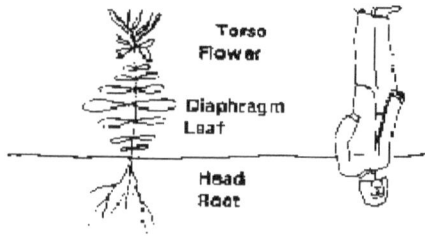

Torso
Flower

Diaphragm
Leaf

Head
Root

For the animal, such as a cow, the situation is much the same as the human however the development of the middle rhythmic system is not so developed, especially in the young animal.

Nerve Sense

Rhythmic

Metabolic

In the past this law was described by the alchemical formula of Sulphur, Mercury and Salt to describe the same threefold nature of physical substances. Salt indicates the act of coming into being through consolidation. These are centripetal forces moving from the periphery to the centre. Astrologys fixed principle. Mercury is the adaptation of the above and below. The leaf shows us this image in its mixing of the water from below with the air from above. Mutable in Astrologese. Sulphur describes the sublimation, dissipation and burning processes involved in the flowering and seeding. These are centrifugal forces moving from the centre outward. Substance is loosened and refined. Cardinal astrologically.

All these pictures provides us with the following associations

MODES	ALCHEMY	SCIENCE	PLANT	HUMAN
Fixed	Sulphur	Thesis	Seed	Nerve Sense
Mutable	Mercury	Synthesis	Leaf	Rhythmic
Cardinal	Salt	Antithesis	Root	Metabolic

Until recently this level of the Biodynamic Vortex was used solely to describe the make up of physical bodies and provide a picture of the 'environment' in which plants grow. There has not been any preparations for this level. However Greg Willis of California has been investigating the use of 'Horn Clay' as a mediating preparation between the activity of Calcium and Silica. He sees this preparation as a mediator between 500 & 501. It should be noted that he suggests this preparation be made from the calcium based bentonite clay.

Level 4

This is the level of the Atmosphere and the Elements.

RS uses this level in several ways. Being primarily the level of the Etheric body, its inner workings describe many aspects of living systems. Life only begins once the etheric body has been internalised. We saw in earlier how a star really only becomes self sustaining once the horizontal plane forms and a 'pulse' develops. (see Astronomical Pictures) So, once the etheric is active, a vast diversity of life forms and processes take place. This four fold model offers us two important insights into how these life processes are organised.

The 'Macro' Polarity was identified as the relationship between the two primary poles - The Cosmic elements of Fire and Air form a polarity to the Earthly elements of Water and Earth. This polarity relationship is based upon the essential relationship between Heaven and Earth, and as such we see it manifesting more in the external processes of nature.

The 'Micro' Polarities are formed when activity comes into play and manifestation begins to take place. Manifestation can only stay active through the interaction of opposites. We saw how the elements of Fire and Earth polarise against each other in the beginning of the great cosmic drama. Once their polarisation and spinning have become intense enough the they suck substance and force into their centre and squirt it out along the horizontal plane. This horizontal plane then

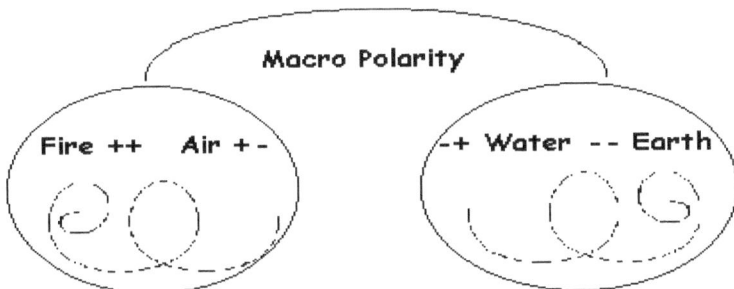

Macro Polarity

Fire ++   Air + -   -+ Water -- Earth

polarises to provide us with the polarity of Water and Air. These are the relationships we see RS describing within living organisms.

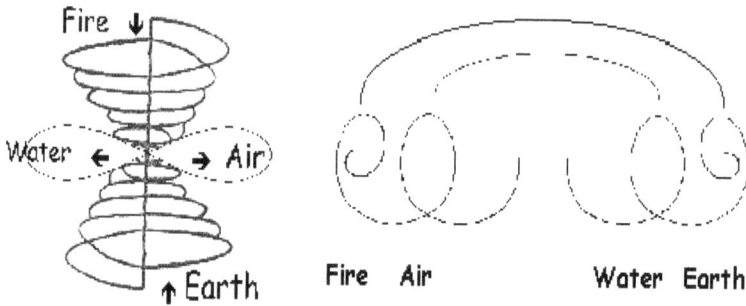

This same process of creation was described by RS in the 3rd lecture of the Agriculture course. In that lecture he describes how the elements of protein H,C,N,O,S work together to form the basis of organic chemistry.

Hydrogen, the carrier of the primary formative force (fire) combines with the basis of physical matter - Carbon (Earth). Together they attract the Astrals carrier - nitrogen (air) which in turn draws in the etheric bodies carrier - oxygen (water) into the organic substances. Sulphur acts as the 'oil' to this process allowing all the other elements to bond and flow around each other.

It has been shown earlier how the four bodies can be associated to the various levels of the spiral and their corresponding 'world' homes. This reference gives the specific outline for the inner workings of each body. If one wishes to consider the physical body there will be three components to consider. Similarly if one is working with the Astral body, there will be seven components to work with and so on. With the Etheric there are four components to consider.

Levels	3	4	5	6
Astrology	Modes	Elements	Planets	Zodiac
Body	Physical	Etheric	Astral	Ego Spirit
'Fold'	3	4	7	12

Before we go into these inner workings it is useful to explore how the four bodies are described by the two polarity structures mentioned above. It is important to clarify the difference between the four bodies working with these two sets of polarities. The Macro polarity has been stated to function on the external manifestations of life, while the Micro polarity indicates the internal workings.

When we look into nature, we can see that the Physical and Etheric bodies work from the Earth and push outwards into space, while the Astral and Spirit bodies move downwards towards the Earth. RS describes this in his medical lectures. (6) This basic process can be seen in plants as well as in Humans. The expression 'feeling low' or 'feeling down', usually used as an expression of physical and psychological tiredness, is an expression of this. The etheric body provides us with a 'watery' cushion between the physical body and the inward moving Astral and Spirit. When the etheric body is strong its uplifting anti gravity effect allows us to have plenty of energy and spark to take on life. Psychological and emotional issues do not weigh so heavily upon us. They are generally able to be dealt with, with a thought or seen in the light of some rational context. As the etheric body becomes drained - lack of food, sleep or water or poisoning by drugs or chemicals etc - the Astral body and the Spirit enter more deeply into the physical body. The etheric cushion does not give us the energy to feed these bodies and so a range of symptoms occur. We can feel tired, run down and look very 'grey'. However the Astrality and ego can act as a stimulant for a time and we start rushing around, like of a coffee high, and start using up our bodies reserves of fat as fuel. This will lead to a massive loss of weight and nervous stress ultimately. Once the Astrality and Ego enter too deeply into the physical organism, psychological and emotional difficulties are not easily moved aside. Obsessions and paranoia sets in. This is also the basis of plant pest attacks as we will see later. What is missing is a strong a buoyant etheric body moving upwards pushing these bodies off the physical. If they enter too deeply illness will occur.

This functioning of the energeic bodies should not be confused with the way RS describes their activity within the living organism. (18) In those lectures he describes associations which are best summarised by the following diagram. It is this diagram we need to refer to, to understand many aspects of RS 'Agriculture'.

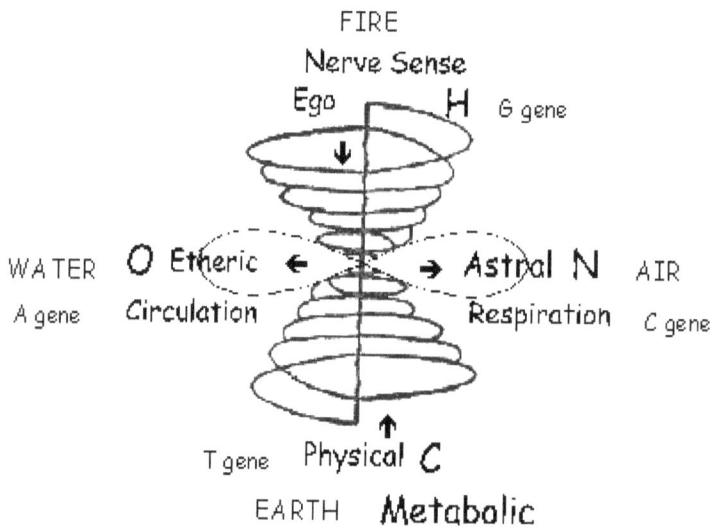

FIRE
Nerve Sense
Ego   H   G gene

WATER   O Etheric ← → Astral N   AIR
A gene   Circulation   Respiration   C gene

T gene   Physical C
EARTH   Metabolic

We need to keep these two separate understandings of the bodies functioning side by side at all times. It is best not to try and mix them up. Be clear when you are using one or the other. Using both simultaneously is a specialised activity.

More 4 fold levels

We have seen in "The Energetic Bodies and the Kingdoms of Nature" how these four levels of creation and bodies work together to form the four kingdoms of nature and how they can be related to the elements.

With each new stage of evolution see 'Evolution on a Pinhead', RS outlines that when each of the energetic bodies came into manifestation - during the Eras - so an ether and its corresponding element also developed. The significance of this is that we are provided with a hierarchy of activity and a set of channels by which any particular energetic bodies activity is carried through into different levels of creation. In 'Creations Layer Cake' we can see the picture of the 12 layers of creation. The outer layers work through the layers below them. RS outlined how Saturns influences will be stronger when it is warm or that the Moon will be stronger when it is wet. (Agriculture) From this we can determine that the condition of the lower levels influence the functioning of the higher ones. Similarly the higher bodies, especially when they do not incarnate, as in the plant kingdom, must use the corresponding parts of the lower levels as their transport into that level of life. Thus when we come to plant growth, the spirit must use Saturns path through the warmth ether, the fire element, cosmic silica, cosmic forces and hydrogen (and the G gene) to make its impact.

So the association of the kingdoms of nature to the elements and ethers, based on the energetic body that separates each level of life, looks like this:

Galaxy	Solar System	Atmosphere	Earth	Spheres
Spirit	**Astral**	**Etheric**	**Physical**	**Bodies**
Will	Psychology	Immunity	Body	**Human**
Nerve Sense	Respiratory	Circulation	Metabolic	**Body Systems**
Warmth	Light	Chemical	Life	**Ethers**
Fire	Air	Water	Earth	**Elements**
Hydrogen	Nitrogen	Oxygen	Carbon	**Biochemistry**
Cos. Forces	Cos Substance	Ter Forces	Ter. Substance	**Phy. Form. Forces**
Cos. Silica	Ter. Silica	Cos. Calcium	Ter. Calcium	**Ca & Si**
Clay	Sand	Humus	Lime	**Soil**
Fruit &Seed	Flower	Leaf	Root	**Plant**
Roundness	Pointed	Wavey	Square	**Forms**
Stalk	Skin	Mass	Tissues	**Plant Growth**
Seed	Ripeness	Size	Quality	**Fruit**
Germ	Seed Coat	Cotyledons	Viability	**Seed**
Nucleus	Mitochondria	Cytoplasm	Cell Tissues	**Cell**
G	A	T	C	**DNA**
North	West	East	South	**Magnetic**

There is one association which becomes important with plant growth. In the two fold workings we were discussing the way Calcium and Silica predominated there activities in plant growth through the vertical and horizontal planes. We can see from the Macro polarity that Calcium has a natural association with the Physical and Etheric bodies activity while the Silica has a natural affinity to the Astral and Spirit workings.

Within the micro polarity however, we see that Calciums working with the horizontal plane brings it into intimate relationship with the

Etheric and Astral activities. While the Silicas relationship to the vertical pole brings it into intimate relationship with the Ego and Physical polarity of activity. This provides us with an indication of another Steiner 'contradiction' which needs to be allowed for when working with life processes.

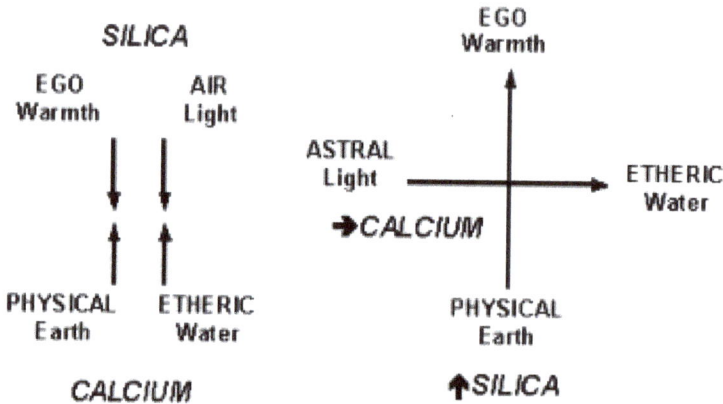

The fourfold level shows itself in Biodynamic often. The differentiation between the Macro and Micro workings of this level is a significant difference to maintain an awareness of. Doing so helps clear up many confusions and 'contradictions'.

SPIRIT Indirect Outer Pl. 1 Warmth Cosmic Forces Clay	Cambium Salamanders ↓	Saturn	P	Valerian	Strengthens the Spirit
	Sylphs	Jupiter	S	Dandelion	Merge Spirit and Astral towards the Physical
ASTRAL Direct Outer Pl. 2 Light Cosmic Substance Sand	Life Sap	Mars	Cl	Nettle	Harmonise the Astrality
		Sun	Ar	Equisetum	Astral stimulates the Etheric within the metabolism
ETHERIC Direct Inner Pl. 1 Chemical Earthly Forces Humus	Undines ↓	Venus	Na	Yarrow	Opens the Etheric to the Astral
	Gnomes Wood Sap	Mercury	Mg	Chamomile	Stimulates the Etheric
PHYSICAL Indirect Inner Pl. 2 Life Earthly Substance Cations	↑ 500	Moon	Al	Oak Bark	Etheric binds to the Physical
		Earth	Si	Quartz	Spirit binds to the Physical

Level 5

The Astrality consists of the 'Macro' Astral seated in the Solar System sphere and the 'Micro' astral body internalised in and around each individual Human. It functions through the archetypal law of the number seven. Wherever you find the number seven expressed you are functioning at an Astral level of manifestation.

As the spirit interacts with life through evolution, the spirit experiences physical sensations as well as emotional psychological and psychic impressions. The Astral body is created out of this activity. Both the impact of the environment on the personal Astral body and the personal experiences of life are accumulated firstly in the personal astral body. Emanations from this can be radiated back outwards and accumulated in the Cosmic Astral body. There is an intimate interplay between both. Primarily though we must accept responsibility for the creation of our own Astral body. It is the qualities of our Astral body which then imprint into and sculpt our Physical and Etheric bodies. It is the working of the Ego\Spirit which hardens this form. The formation of the Astral body can be seen as similar to the formation of the planets around a Sun. Once the Sun reaches a certain intensification in its centre it excretes matter along the horizontal axis. This matter then organises itself according to the Solar electro magnetic fields. Due to the spinning motion of the solar system this matter is then accumulated into balls, which become planets. Similarly our spirit accumulates astral clutter.

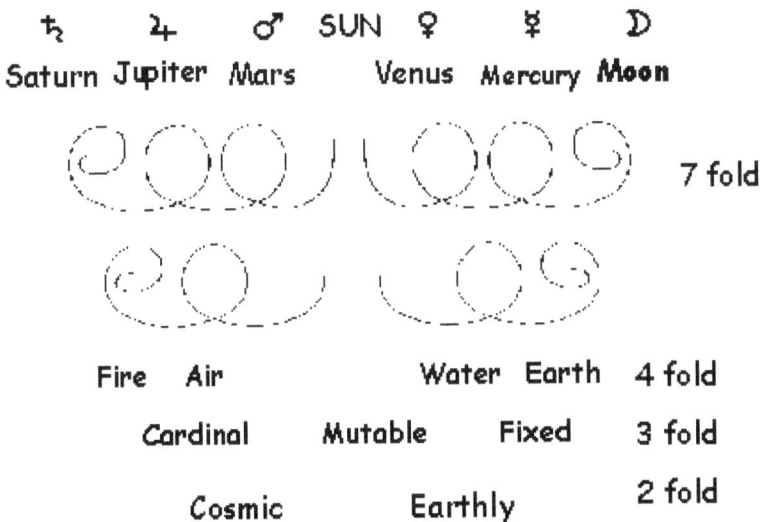

♄	♃	♂	SUN	♀	☿	☽	
Saturn	Jupiter	Mars		Venus	Mercury	Moon	

 7 fold

Fire	Air		Water	Earth	4 fold
	Cardinal	Mutable		Fixed	3 fold
		Cosmic	Earthly		2 fold

The Astral body of the human equates to our personality, instincts, desires, fears, and is certainly a place our emotions are played out. Emotions in particular arise as an interplay between the Etheric body and the Astral.

Upon death we drop our physical and etheric bodies. However our astral body and spirit then have some talking to do as to what happens next.

Just as we found in the level four activity, the planets also divide themselves into a Macro division, which describes external laws and a Micro division which describes internalised laws, manifest due to the polarisation of the planets

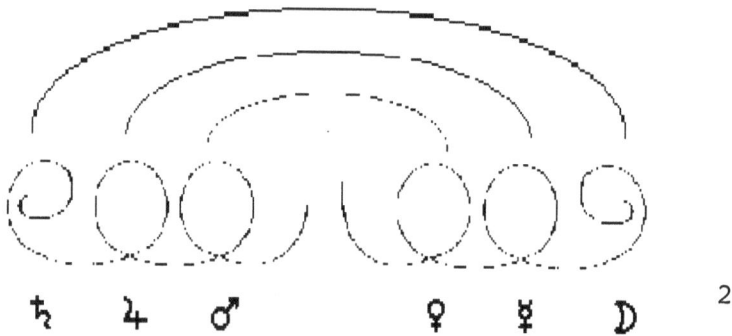

♄   ♃   ♂   ♀   ☿   ☽   2

On diagram 2 we have the micro polarities

In diagram 3 - 1 is the micro polarities

 2 is the planets lined out arranged according to the speeds of their rotation within the Solar system

3 is the Macro polarites of the inner and outer planets.

All three structures are relevant and used by Steiner. He also used another order where the Venus and Mercury positions of 2 were exchanged. This is an order related to the Earths experience of the solar system. It is a more subjective view of existence than No 2 presented here.

The seven planets of the Solar system act as mirrors or adjusters to the formative forces coming from the Fixed Stars. These streams of forces come into existence in various forms. The organ systems of living entities are a result of these forces coming into manifestation. The organs formation and continuing action, in a sense, brings these planetary forces to an end in the physical body. Not all animals manifest these organs internally to the degree the Human does. In fact the degree by which organs are manifested indicates the degree the Astral works into the physical \ Etheric organism. (21)

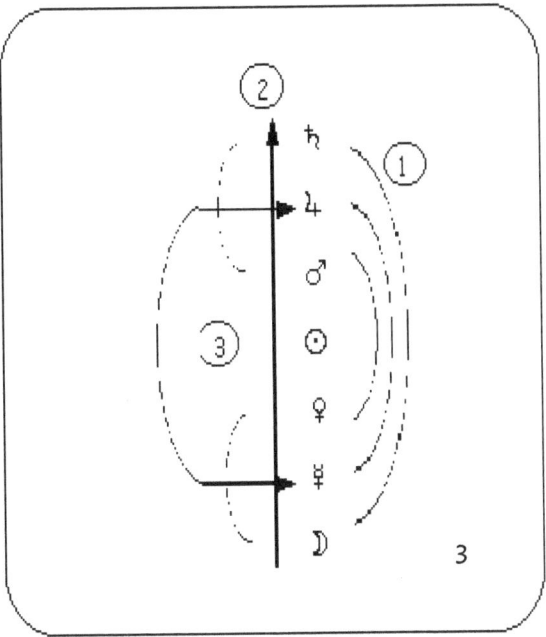

The general correspondences are as follows. In each series the planetary polarities of Saturn \ Moon, Jupiter \ Mercury, Mars \ Venus all work.

The physical ' organs' are associated thus

Saturn - Spleen	**Jupiter - Liver**	**Mars -- Gall Bladder**
	Sun - Heart	
Moon - Reproduction	**Mercury- Intestines**	**Venus -Kidneys**

The Metals

Saturn - Lead	Jupiter - Tin	Mars - Iron
	Sun - Gold	
Moon - Silver	Mercury - Mercury	Venus - Copper

Within the Etheric bodies action and its relationship to the elements the planets work in the following way:

109

| Saturn - Warmth | Jupiter - Light | Mars - Dry |
| Moon - Cold | Mercury - Dark | Venus - Wet |

The chakra system of energy is one of the main Human manifestations of the Astral body. The chakras act as the organs of the Astral body. These are energy centres acknowledged by most energetic traditions and are easily experienced. The functioning of these chakras can be associated to the 'health' of the organs.

Saturn - Crown	Jupiter - Third Eye	Mars - Throat
	Sun - Heart	
Moon - Base Plexus.	Mercury - Sacral	Venus - Solar

Psychologically the social development work has identified seven personality types

Saturn - Investigator	Jupiter - ThinkerMars - Entrepreneur	
	Sun - Balanced being	
Moon - Conserver	Mercury - Innovator	Venus - Carer

A further investigation of the planets can be found in 'Astrological Science' and 'Spiral Astrology'

Planets and Plant Form

RS outlined plant form as being primarily influenced by the working of Calcium and Silica substances and forces. He outlined how the outer planetary influences of Saturn, Jupiter and Mars were carried into activity along the Silica pathways. While the inner planetary influences of Moon, Mercury and Venus are carried into activity along the Calcium pathways. Lievegeods book looks more deeply into the workings of these planetary streams and their effect on plant growth. In keeping with what was outlined earlier about the primary and secondary Calcium and Silica streams, Lievegoed also suggests that the planets also have primary and secondary activities. This is a complex picture which will be looked into further in the "Biodynamic Plant Growth" chapter. We can start to make an approach to understanding this picture once we have looked into the 12 fold organisation of creation. For now this diagram will have to suffice.

Planet Archetypes in Plant Growth

The terms inner planets and outer planets holds the key. There is a broad division between trees , that harden wood and hence live longer than one year, and plants that remain annuals or as soft herbacous plants. Within these broader definitions the individual planets can be related to specific plant areas.

The outer planets all have cycles longer than one year. Mars-2 years, Jupiter -12 years, and Saturn -30 years.

The inner planets cycles are all shorter than one year. Venus-256 days, Mercury 100 days and the Moon-28 days.

Outer Planets

Saturn rules the evergreen trees and especially the conifer varieties. Their rhythms are usually of long duration and their lifespans are in multiples of 30 years. The restrictive quality of Saturn is noted in their pyramidal shape. Their sap is slow and often resin like.

Jupiter trees are deciduous and form spherical canopies, with their sap being more fluid and their lifespan usually of moderate duration. Fruit trees and the Oak family are good examples of Jupiter. The abundance of Jupiter is evident in the fruit we gain from her charges.

Mars' cycle is much shorter and governs both biennial plants and the shrubs, especially the shrubs that have an element of die back in their growth. After flowering and fruiting the new shoots come from the growth below.

The Inner Planets

Venus rules plants in which flowering predominates and has been associated to the Alpine and desert plants. These plants lie dormant for long periods, and when the conditions are right, flowering occurs. Their flowering can be extremely short, but is the highlight of their life.

Mercury runs and weaves following the path it is pushed. The plants it governs are the vines and the runners. Mercury's mark is seen in any climber that does not support itself.

The Moon has the shortest cycle and is associated with quick growing plants, and those that retain a large content of water. On the one hand, this governs Cacti and succulents, and on the other it governs most of our vegetable plants.

The Sun is the mediator of all the planetary forces and its plants show a balance of all the parts. The clovers and grasses are Sun ruled. The roots, leaves and flowers all have equal predominance in the life cycle.

These divisions give some hint as to why planting by the Moon is so popular among vegetable growers, as the Moon has the closest affinity with vegetables. Experience has also shown that emphasising the influence of other planets at the sowing, planting and harvesting of their associated plants, helps in their growth and vigour.

These associations are archetypal in nature and while the varieties mentioned, carry the mark of the planet most prominently, there are many trees that carry a mixture of influences.

Another sevenfold division of the plant kingdom has been made, according to the archetypal evolutionary stages, outlined in "Evolution on a Pinhead".

Saturn	Fungi
Sun	Algae
Moon	Mosses
Earth	Ferns
Mercury	Gymnosperm
Jupiter	Monocotyledon
Venus	Dicotyledons

The Compost Preparations.

The compost preparations are where we find the planets activity most important. Dr Steiner outlined in some detail a series of six preparations. These, he suggested, should be placed in the compost heap. Their overall activity is to help balance the energetic bodies' activities in the earth, plant, animal and human. While they aid and regulate the composting process, their energeticly balancing activity, is carried to the soil through the medium of the compost. In more recent times they have been inserted into small manure pits and applied directly to the soil. The information he supplied regarding these preparations was quite comprehensive and came in two parts. The first part was in lecture 3 where he covered the elements of protein, - the basic molecule of life. Here Rudolf Steiner suggests that the four 'bodies ' are carried into manifestation by the everyday elements of Oxygen, Nitrogen, Hydrogen and Carbon, aided by Sulphur, although in this lecture it is not as clearly stated as in the medical lectures (see - Hydrogen the carrier of the Ego)

		(Saturn	Valerian
Positive Sun	Outer Planets	(Jupiter	Dandelion
Silica		(Mars	Nettle
501			
		+ Sun	Silica
		- Sun	Manure
500		(Venus	Yarrow
Negative Sun	Inner Planets	(Mercury	Chamomile
Manure		(Moon	Oak Bark

As we have already seen, he suggests The Ego is carried by Hydrogen; the Astral by Nitrogen ;The Etheric body by Oxygen; and the Physical by Carbon. He then adds that the Elements of Sulphur and Phosphorus are carrier substances that help these four to blend and intermingle.

The significance of this information, is possibly not recognised until the fifth lecture, when Dr Steiner outlined the Bio-Dynamic 'compost' preparations. In this talk he described the making procedure of each preparation and outlined a sketch of their activity. At first glance these preparations appear unusual, as they are comprised of herbs, put (generally) into animal organs and then placed in the ground during specific periods of the year.

The general combinations are:

Prep	Herb	Receptacle	Period
507	Valerian	Water	Immediately
506	Dandelion	Mesentery	In Earth over winter
504	Nettle	Earth	In Earth all year
502	Yarrow	Stags Bladder	In Air/Summer, Earth/Winter
503	Chamomile	Intestines	In Earth for Winter
505	Oak Bark	Skull	Water for Autumn and Winter

The numbers were given in the order that Steiner outlined them. (They were originally classified and thus numbered as homeopathic medicine.) On inspection these preparations follow the standard line of thought of basic sympathetic magic, being that if one combines substances of similar activity from different levels of life, one creates a result greater than the sum of the parts.

The easiest examples are the preparations made from yarrow and Chamomile. Both of these herbs are widely recognised to aid the healing of the organs they are placed in.

Further associations with these herbs, to their sheaths and their planetary correspondences, has been dealt with in a fine book by Dr B. Lievegoed. Titled "The Working of the Planets and the Life processes in Man and Earth", this is a thorough study of the preparations, their activity and their relationship to the activity of the planets. (see also (7))

From the Astrological Model, this information corresponds to the level of the planets and utilises the number seven. The Sun is the harmonising central influence of the solar system and so it is in this level as well. It can be seen as the centralising and manifesting form of life itself harmonising the polaric interplay's of the other planets. The Sun preparations (500 & 501) act as the overall 'parents' of the other six compost preparations.

B. Lievegeods overall associations are as follows. Also included are the chemical elements Dr Steiner outlined to each preparation.

Prep	Herb	Planet	Element
507	Valerian	Saturn	Phosphorus
506	Dandelion	Jupiter	Hydrogen
504	Nettle	Mars	Nitrogen
502	Yarrow	Venus	Sulphur
503	Chamomile	Mercury	Oxygen
505	Oak Bark	Moon	Carbon

It should be noted that the planetary order that arises from this model of Lievegeods is the same as that in the Astrological Model. Further to this Lievegoed outlines that the preparations are in polaric balance to each other in the same way that has been outlined in my earlier discussions regarding Level 5.

North

	Secondary planets	Primary planets	
C O S M I C	Seed ♄ ♑	♒ ♄	Archetype
	Etheric Oils ♃ ♐	♓ ♃	Plastic Forces
	Protein ♂ ♏	♈ ♂	Growth in Space
	West ↑	↓ East	
E A R T H L Y	Excretion ♀ ♎	♉ ♀	Nutrition
	Supporting Organs ☿ ♍	♊ ☿	Sap Flow
	Tissues ☽ ♌	♋ ☽	Reproduction

South

Manifestation - 2	Being - 1
Chemistry-Physical	Antichemistry - Ethereal
Destruction Stream	Upbuilding Stream
Visible Formation	Invisible - Dynamic
"Substance"	**"Forces"**

Dr Lievegoed

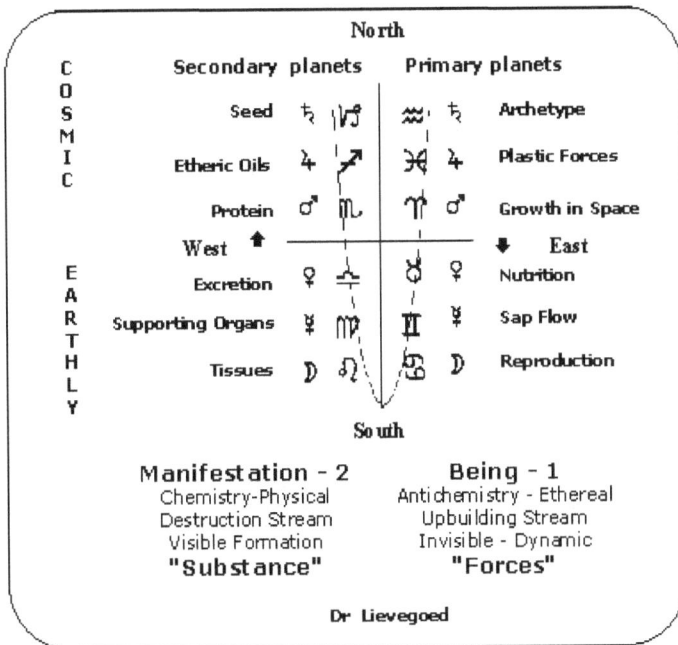

As just mentioned, the Sun acts as the central balance for the planets. These preparations can be therefore separated into the inner planet group and the outer planet group, giving us the following association.

The Outer planet preparations can be seen as more subtle variations of the Silica preparations activity. Hence working with nutrition and seed formation. While the inner planet preparations, as three subtle variations of the activity of the Manure preparations, enhance the processes associated with reproduction and growth.

Since this book was first published, I have become aware of understandings most prevalent in the USA, that provides a different association of the preparations to the planets. This association, outlined by Hugh Lovel, includes the preparation RS outlined made from Equiseteum. Upon investigation I am inclined to offer a variation upon the system Hugh puts forward.

Lievegeod	Atkinson	Planet	Lovel
507	Valerian	Saturn	Equiseteum
506	Dandelion	Jupiter	Dandelion
504	Nettle	Mars	Valerian
500&501	Equisetum	Sun	Nettle
502	Yarrow	Venus	Yarrow
503	Chamomile	Mercury	Chamomile
505	Oak Bark	Moon	Oak Bark

The Preparations

Most BD practitioners generally see the compost preparations as being used as soil and growth regulators. With a vague sense they work on more subtle levels as well. Most recently their activity was publicly described as being due to their bacterial and enzyme content, which arose from a bacterial reaction between the internal substance and the sheath into which it was placed. Sadly there appears to be only a small conscious appreciation of them as substances which actively work with the energetic as well as physical realms of the land and plants.

Some BD practitioners accept that healthy life is created through a balanced interplay of the various energetic bodies. However they have little understanding of how this is achieved. The basic assumption is that this is brought about by the spraying of preparations 500 and 501 as well as including the compost preparations in the manuring cycle. But even with this activity there are still situations , especially in the changeover phase, where imbalances are evident for some years.

The question arises 'Could these preparations be used in a more active manner to correct specific examples of imbalance?'

The answer is YES, however it is a rather complex process, calling on the exponent to have a thorough experience of each preparation's activity, their influence on the energetic bodies and the energetic bodies' activity in the living organism, be it plant, animal or human. Plants rapidly come into situations in which various bodies are imbalanced. Imbalance is caused by soil quality, mineral imbalance, weather disturbance and general neglect. They all work simultaneously and at different times and combinations. Nature is a living moving organism. The BD preps offers us a chance to moderate these swings into balance, which provides the difference between the success and failure of a crop.

The Horn Manure and Silica preparations can be seen as broad stroke preparations that powerfully stimulate the two primary formative growth processes. Their activities are easily seen, and at times, work too broadly to be used as specific remedies. They each act directly on enhancing one pole of the plant's activity or the other. This can inhibit their ability to be used frequently.

One of the basic premise of Biodynamics is that pests and disease are an imbalance in the workings of the energetic bodies. Therefore to remove them one must remove the conditions which allow for the problem to be there in the first place. Hugh Lovel recently awakened me to **"with the BD preps we are organising otherwise unorganised energy"**. I have often imagined health as the four bodies being more or less integrated into each other. In harmony their integration is such that there are no 'holes' or 'vacuums' in them. As physical conditions deteriorate so the bodies being to separate. Vacuum holes open up and insects and disease are literally sucked into these holes. Once the holes are closed again the insects just go away. I have experienced this 'controlled' coming and going phenomena many times.

When using preparations as remedies, one attempts to apply sufficient influence to the environment, that the conditions fostering the growth of the pest or disease, no longer exist. While 500 & 501 do this, they also carry with them the power to unnecessarily activate only one area of plant growth, so symptoms of excessive use easily arise. In the case of 501 sunburn easily occurs as does the tendency for plants to take on the stature of the dry landscape or go off to seed. Leaves thicken and become waxy, and in some instances, plants that normally have leaves in a downward facing nature, stand upright.

500 in excess encourages leaf development and inhibits the flowering tendency of a plant.

These preparations tend to encourage which ever pole they govern and not to harmonise and balance the activity of the energetic bodies. So their use, especially 501, should be made with discretion.

The 'compost' preparations act as a "second tier" of the same activity. From the model it is apparent that the 'outer planet' preparations are associated to the Silica prep, and the 'inner planet' preparations work with the Manure prep. The compost preps work more specifically than the 500 & 501 and are softer in their action.

They appear to do the job of bringing any specific activity back into the right relationship with the others, without affecting the essential nature of the plant, so these preparations can be considered tools of dynamic husbandry. The decision to include 500 or 501 will depend on the severity of the problem.

Specialisation of these preparations can be achieved by their potentisation. However this is an area of some exactitude and should not be attempted by the novice. Imbalance easily occurs with prolonged and indiscriminate use of single potencies.

One of the limitations to the preparations being used as single preparations and as energetic body balancers has been that there is no real work into establishing their independent functions. Also a science of energetic activity has not been fostered within the Biodynamic movement over the last 100 years. Part of this is due to an inability of the Agriculture course to be understood in depth and discussed in a manner which allows for concrete conclusions to be drawn. This is where I believe the Astrological model can solve these problems. By providing a frame of reference for Biodynamic ideas and methodologies it is possible to see how one piece of information fits in with other pieces so that some conclusions can be made. If nothing else as a basis for further practical experimentation.

While giving the making process of the preparations, RS also outlined the basic effects of each preparation. He also gave some of the chemical processes that occur in each. In two cases, he indicated the influence a preparation has on the energetic bodies. However, he did not extend this to all of them. As this is an important aspect of the overall thrust of this work, I have made an initial attempt at

Saturn	Jupiter	Mars	Venus	Mercury	Moon	5
507	506	504	502	503	505	

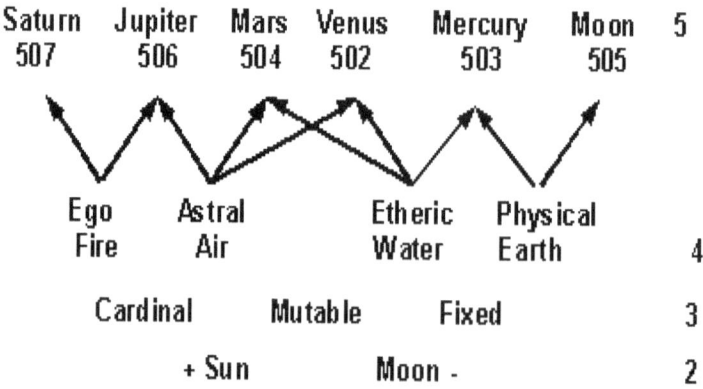

Ego	Astral		Etheric	Physical	
Fire	Air		Water	Earth	4

| Cardinal | Mutable | Fixed | 3 |

| + Sun | Moon - | 2 |

extending the interpretation of each preparation to include their energetic activity. A summary follows:

(It seems appropriate in this section to quote from Steiner directly. He was born with the Sun in the sign of Pisces with Mercury, the planet of communication conjunct Neptune, the planet of images. His method of communication, as you will see is very pictorial and imaginative.)

We have seen already how one level of activity can be associated to higher and lower levels. With the information we have accumulated already it is possible to produce the 'Biodynamics Decoded' diagram.

This diagram shows the vertical movement up the Astrological spiral with the Biodynamic cross references added. It is apparent from these associations (and the comments made by Dr Steiner and B Lievegoed) that conclusions can be immediately drawn as to what the complete set of energetic workings of the compost preparations might be. It is apparent that the outer planets preparations will most likely work on organising the Astrality and Spirit functions while the inner planet preparations would most likely organise the processes of the etheric and physical bodies activity. The middle preparations would tend to harmonise the interplay of the two overall spheres.

The Outer Planet Preparations

507 - Valerian - Saturn - Tincture.

All that was said about this preparation was that it will stimulate the plant "to behave in the right way in relation to the phosphoric substance."

This is a seemingly vague comment, but seen in relation to comments regarding the other preparations, and seen in the context of the Astrological model several hints arise. Firstly being the Saturn preparation it has a relationship to the Fire/Ego pole. Phosphors as an element is active in the strengthening of the nervous system processes within living organisms, which is in turn a manifestation of the Spirit's activity. In plants it is credited with creating strong root systems - the nerve sense pole. Valerian is a sedative and calmer of the nerve system. 507 has long been known as a warming preparation and is used for frost protection. 507 is also seen as a cover over the compost heap helping to contain the activity of the heap inside itself. In lecture 8 where RS talks of human nutrition we are presented with an image of the Spirit's process (called Cosmic Forces) and how they act to cap or reflect back into the organism, that has been rayed out to them from the Earthly Matter or terrestrial calcium, liberated from the digestive tract.

In "the Anthroposophical Approach to Medicine" by Hausemann & Wolff we are told phosphorus 'establishes "the connection of the ego to the substances of the body." "The astral body lives in opposites that can be best be seen in the concepts of sympathy and antipathy....hyper and hypopthyroidism etc. The ego has the task of standing in between them, but not succumbing to either" " In working from imbalanced states toward the development of those that are balanced the ego requires phosphorus"

All this information leads to the conclusion is that the Valerian preparation **Strengthens the Ego against too strong an Astral activity**. It works with the element of Phosphorus.

506 - Dandelion - Jupiter - Mesentery - Soil for the Winter

This preparation was discussed in length bringing several interesting facts to light. Some of this discussion is alchemical in nature, and scientific minds I have discussed it with, find the transmutation of elements outlined here difficult to understand.

Dr Steiner says there is " a quantitative relationship between Limestone (Calcium Carbonate) and Hydrogen, similar to that of oxygen and nitrogen". He goes on to say "that under the influence of Hydrogen, Limestone and potash are transformed into something like Nitrogen."

To gain some clarity on this statement we can remember that at level four, we talked of the micro polarities that existed in the elements.

Fire - Ego - Hydrogen   }
Air - Astral - Nitrogen   ) }
Water - Etheric - Oxygen) }
Earth - Physical - Carbon   }

Here we see the same micro polarities occurring.

R.S. goes on to say "that the oxygen/nitrogen process occurs in the air, while the Hydrogen/Limestone (Carbon) process occurs in the organic process." He does not enlarge on this element "something like nitrogen" but goes on to discuss the transmutation of silicon in the plant into a substance not yet included in the chemical elements. Silicic acid is needed to attract the "cosmic influences" to the plant and this dandelion preparation "mediates between the silicic acid in the cosmos and that which is needed in a given district."

This preparation is " thoroughly saturated with cosmic influences. He suggests the plants will become "sentient to all that is at work in their environment" so they can attract to themselves what they need."

From the Astrological model this is the Jupiter (mutable) preparation and so is sitting in between the elements of Fire and Air. This suggests that some harmonisation of the ego and Astral bodies will be involved in its action.

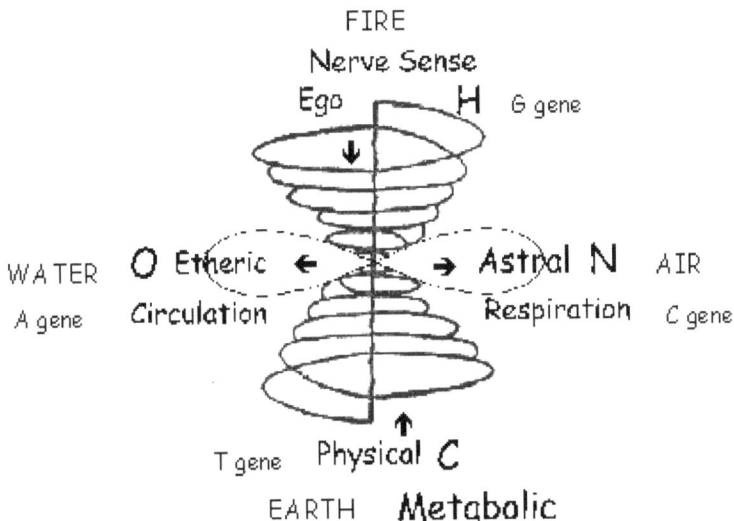

FIRE
Nerve Sense
Ego   H   G gene

WATER   O Etheric ←   → Astral N   AIR
A gene   Circulation   Respiration   C gene

T gene   Physical C
EARTH   Metabolic

121

From the two references he uses, Hydrogen-Carbon & Silicic acid mediation, we are left with the image of bringing heaven and earth to meet. More specifically, the elemental diagram outlined above would suggest this preparation will help the **Ego (& Astral) to enter more strongly into the physical realm.**

504 - Nettle - Mars - In the Earth - All year

From the astrological model, 504s relationship to Mars places it in an intermediatary position between the elements of Air and Water. Hence we can surmise that it works in some way between the Astrality and the etheric body, with an emphasis on the Astrality side of the equation.

The nettle preparation follows more conventional lines of thought. Chemically R.S. suggests this preparation incorporates sulphur, potassium, and calcium with a kind of iron radiation. Stinging nettle is known for its ability to concentrate elemental iron and is often used in early spring in soup as a tonic and blood purifier. Iron is the element of Mars in traditional association. Dried blood is extremely high in nitrogen content. In lecture three R.S. discusses the role of nitrogen as the carrier of the Astral and suggests it is "a very clever fellow". Again in the description of this preparation he uses the same images. Suggesting that this preparation allows " the Manure to become inwardly sensitive, we might say" "It will permeate the soil with reason and intelligence" that " will not suffer any undue decomposition's to take place, and no improper loss of nitrogen."

Later it is suggested that "The soil will individualise itself in a nice relationship to the particular plants in the environment"

From these descriptions we can conclude that the preparation harmonises and strengthens the working of Nitrogen and likewise the working of the astral in the environment, enabling the Astral to find its proper place. Through its activity of individualising the soil it is suggested that it **encourages the harmonious workings of the Astral**

These are the three outer planet preparations and, as can be seen, their activity is primarily focused on the balanced interplay of the Ego and Astral, in keeping with the overall activity of "501", the Silica preparation, and with the overall associations of the Astrological Model.

The Inner Planet Preparations

502 - Yarrow - Venus - Stag Bladder - 6 months in Air/6 months in ground

R.S. described this preparations' activity as "bringing sulphur into the right relationship to the other substances" and "correcting all that is due to the weakness of the Astral body"

Further he suggests it " reendows the manure to quicken the Earth (so) that the more cosmic substances of silicic acid and lead are caught and received."

Again we are given the picture of the astral body's interaction with the plant, but this time the preparation is aiding where the Astral body is weak and can not take a proper hold on the Etheric and Physical. Lievegoed makes the observation when talking of the invisible primary process of Venus, (it) " Opens the Etheric Formative Forces into a cup or chalice and nourishes what Mars thrusts into space."

Being in the inner planet polarity it would suggest that this preparation **opens the Etheric to receive the Astral.**

503 - Chamomile - Mercury - Intestines -In the ground through the winter

This preparation is described as "Helping bind the calcium substances to receive life to itself and transmit it to the environment."

The use of the word 'life' here is a reference to the Life body or the Etheric. So this preparation strengthens the Etheric body. Later R.S. suggests "It assimilates that which can chiefly help to exclude from the plant those harmful effects of fructification." Some light can be shone on the statement 'harmful affects of fructification' by looking elsewhere in the course,(lecture 6) to where R.S. was discussing the true nature of fungal attack. He suggested that fungus was caused by the processes of fruiting occurring at a much earlier stage of development than it otherwise should. From our model it can be seen that flowering and fruiting are activities of the Astral and Ego. So he is suggesting that this preparation will hold back the activity of the Astral and Ego if they occur too early by strengthening the Etheric body.

In summary this preparation **Strengthens the Etheric against the Astral (& Ego)**

505 - Oak Bark - Moon - In skull - All year

R.S. was fairly straightforward with this preparation. He suggested " by bringing calcium into the soil it restores order when the ether body is working to strongly, that is when the astral cannot gain access to the organic entity. It dampens down the Etheric."

He describes the process as follows. " We use the bark of the Oak so that the shrinking is beautiful and regular and does not give rise to shocks in the organic life." More directly he suggested that this preparation "will lend forces prophylactically to combat or arrest any harmful plant diseases." Elsewhere RS commented that the bark of a tree should be seen as turned up Earth and that the stalks and leaves can be seen as small plants growing out of this earth.

In summary, **it retards a rampant Etheric body by sucking it more tightly to the physical.**

These inner planet preparations are concerned primarily with the activity of the Etheric and Physical bodies, helping them to find their proper place with regards the Astral and Ego, assisting the Horn Manure preparation 500 in its overall activity of harmonising the terrestrial sphere.

Putting the above information all together we have the following:

Practical experience has shown these preparations can bring balance between any irregularities in the working of the bodies.

I regard these preparations and the accompanying information as a real gift to humanity from Rudolf Steiner. They give the farmer and gardener access to a simple, safe tool kit that can heal and balance their environments. Using the view that all disease and pest attack, in plants and animal, are a result of an imbalance occurring in their physical/energetic balance, we now have eight substances at our disposal that can consciously influence the way these bodies interplay and so a starting place for finding sensitive, purely BD remedies for pest and disease control.

These preparations are generally inserted into a compost heap or liquid manure to balance the raw Physical, Etheric, Astral and Ego forces released during these decomposition processes.

Furthermore these preparations could be used individually and potentised homeopathically to further specialise their action, to harmonise imbalances as they occur during a growing period.

It is best for the novice to use the original preparation in addition to their sprays of 500 and 501 until experience is compounded. However a healthy sense of exploration and experimentation is certainly encouraged.

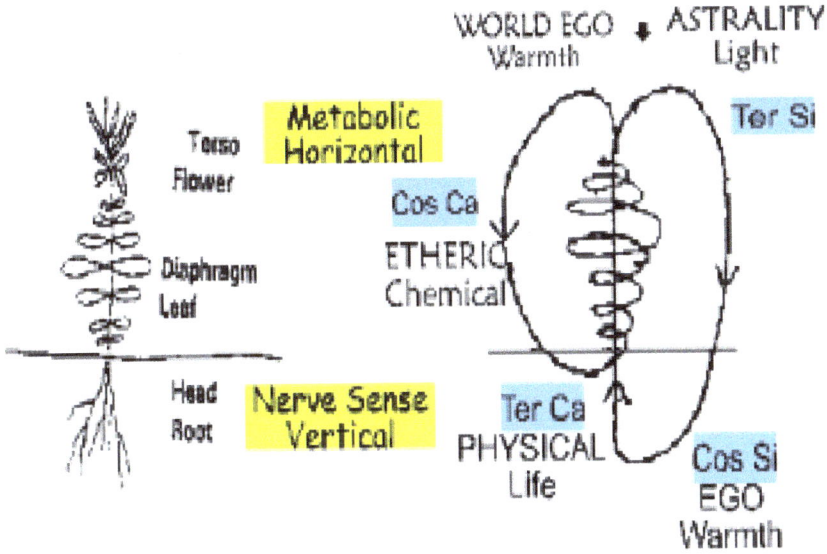

Level 6

The Ego sphere functions from the World Ego placed in the Galaxy and through the personalised Ego, which is incarnated in Humans and functioning externally as the group soul and Devas in the animal, plant and mineral kingdom. While the Ego is not incarnated in these other spheres its presence is seen. The Ego sphere is the source of the Formative Forces, hence it is the repository of all the archetypal forms of each species of life on this planet. It could be surmised that each species of plant and animal has a guiding star of which it is the manifestation. Whether it is this simple I would question. While life forms are a mixture of many influences, the archetypal impulse which keeps a rose , a rose comes from the fixed stars. The variation in form,colour and variety is influenced by the planets, the environment and weather.

The basic patterning of the Ego sphere is carried by the zodiac. Two zodiacs are used, Steiner and his followers use the usual Aries through Pisces zodiac for many applications. However RS has also presented a lesser known Cancer through Leo zodiac, which runs in the opposite direction.

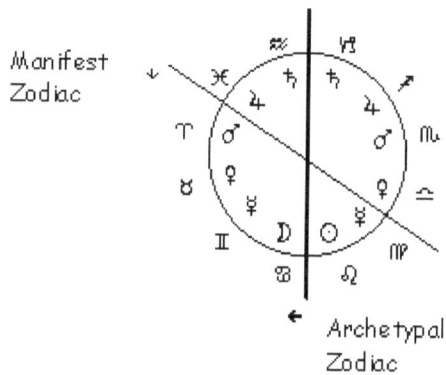

Manifest Zodiac

Archetypal Zodiac

The basic structure of this level can be found in the Astrological model

This region represents the Cosmic spirit sphere, hence the number twelve will indicate the energetic seed or Egoic aspect of a given subject, wherever it can be identified. This twelve folding outlined by the zodiac can be carried over as an archetypal structure into many areas of life.

Mineral Kingdom - Chemistry

Rudolf Hauschka outlined in " The Nature of Substance " his appreciation of how the Zodiac forces work into the chemical sphere. The editorial of this work state that this is a hypothesis, and some of his trials have not been able to be recreated. However, there is still considerable merit to the work, so they have continued publication. His conclusions are compelling and worth studying. Through excellent observation of the elements concerned he has developed an imaginative association with the constellations of the zodiac.

There is one area with which I have great difficulty with his thesis and its a limitation shared by many northern hemisphere writers. Dr Hauschka associates the qualities of the elements to the constellations through their placements within the seasons of the year of the northern hemisphere. Seasons are conditional as to which hemisphere you are in. So his summer constellations become winter constellations within the southern hemisphere. This is a very limited view of the world which has, more than once, been the sole reason for totally invalidating Astrological arguments within the scientific community. This association is inappropriate and indeed IMHO incorrect.

His relationship to the constellations are very convincing nevertheless. Note he is using a zodiac which begins with Leo and runs backwards to Virgo

His associations are

Leo	Hydrogen
Cancer	Phosphorus
Gemini	Sulphur
Taurus	Nitrogen
Aries	Silica
Pisces	Chlorine
Aquarius	Oxygen
Capricorn	Aluminium
Sagittarius	Magnesia
Scorpio	Carbon
Libra	Calcium
Virgo	Sodium

Another contribution he has made was to point out that the elements of protein primarily exist within the atmosphere - Air realm. He goes on to outline how the fourfoldness shown by these elements carries over into the Hydrosphere - the water realm and its main constituents, Sulphur, Halogens (F, Cl), Magnesia and the Alkalis (K, Na). Also in the geosphere-the mineral realms constituents, Phosphorus, Silica, Aluminium and Lime.

	Atmosphere Cardinal	Hydrosphere Mutable	Geosphere Fixed
Fire	Hydrogen	Sulphur	Phosphorus
Air	Nitrogen	Halogens	Silica
Water	Oxygen	Magnesia	Aluminium
Earth	Carbon	Alkalis	Lime

Interestingly, if we looked at the Atmosphere are the Cardinal pole, Hydrosphere as Mutable and Geosphere as the Fixed pole we come up with a different cross reference for constellational influence.

The Plant Kingdom

To date no real 12 fold division of the plant kingdom has come to light. There have however been developments in the way plants are influenced by the zodiac in their growing habits

The Planting Calendar

While the Zodiac is referred to in general Anthroposophical literature , the most common use in Bio-Dynamics for the zodiac, is centred around the 'Planting' calendar. Based on work by Maria Thun and Agnes Fyfe, the Bio-Dynamic community has developed an extensive system of planting by the moon. It should be mentioned that the central authority for Bio Dynamics in Europe has issued articles saying they cannot support Maria Thuns findings. They have been unable to reproduce her trials in the many years they have been trying. The work of Agnes Fyfe still stands to be challenged.

The system is based on the five Moon rhythms and their cycles through the constellations of the zodiac. This Moon data has become the basic framework for the timing of Bio-Dynamic activity. As part of targeting the effectiveness of an activity, it is important to stir the preparations or sow the seed at a cosmically appropriate time. The B.D., planting guide - based on the Moon rhythms - has become the device by which these times are identified.

Apparently, Thun and Fyfe did not use Astrological reference in their experimentation. However, their findings are 'Astrologically' correct. The experiments in Thuns case were/are carried out by sowing seeds everyday, watching their growth patterns, and then in retrospect, observing where the moon (and other planets) were placed. Fyfes' experiments were different in that they included the use of Chromatography pictures of the plant sap at the time of harvest. (17)

Both researchers came to the same conclusion. The moon's movement through the unequal constellations of the zodiac - as they appear in the sky- had a definable influence on plant growth. More specifically, they proved that the element (level 4) of the constellation, in which the moon was placed, was enhanced by its presence, i.e. it has a direct bearing on the corresponding crop type. This means that all fire constellations influence the Fruiting plants and the water constellations influence the Leaf and foliage plants etc. There is however no differentiation made between the constellations of the same group.

The influences of the other planets have also been defined but these are not generally used in common B.D. practice.

A general summary of the principles of Moon Planting.

The five moon rhythms all influence the plant in a manner similar to the Sun in its annual journey around the Earth. The Sun's path can be broadly divided into two stages. From Mid winter to Mid-Summer and then From Mid-Summer to Mid-winter. During these times we see an expansion of plant life from the winter solstice to the Summer solstice. Similarity plant life moves through a contracting phase from the Summer solstice to the winter solstice.

Moon rhythms also influence the plant towards expanding or contracting.

The Moon rhythms are: Waxing and waning - the lit surface of the Moon enlarges and diminishes according to its angle with the Sun.

Ascending and Descending - As the Moon moves around the Earth, its daily path rises higher for two weeks, then lower across the sky. In the same way the Suns' path appears too through the seasons of the year.

Nodal Rhythm -The nodes are the points where the paths of the Sun and the moon cross. These points revolve around the zodiac in 18.6 years in a retrograde direction.

Apogee and Perigee -Due to the Moons path being an ellipse, it moves close to, and away from the Earth. Apogee is the furthest point. While Perigee is the point closest to the Earth.

Through the Constellation -The constellations are the groups of stars in front of which all the planets move. The elements that relate to the particular constellation connect them to certain parts of the plant - Fire/Seed, Air/ Flower, Water/Leaf, Earth/Root. (level 4)

While these are all individual rhythms and act independently of each other, they all have an expanding and contracting phase. The Expansion phase is an out breath from the Earth and draws sap into the upper parts of the plant, enlivening the flowers. The contracting phase is an in breathing towards the Earth and draws the plant sap towards the Earth.

These extremes can be are related to Level 2, the expansion to the Male Sun and the contraction rhythms to the Female Sun.

The two extremes can be listed as follows:

	Expansion	Contraction
Sun	Winter Sol. -> Summer Sol.	Summer Sol. -> Winter Sol.
Moon	Waxing period	Waning period
	Ascending period	Descending period
	Apogee	Perigee
	Air and Fire constell.	Water and Earth Constell.
	Ascending Node	Descending Node

The phases of the moon and the ascending / descending periods are used in a similar manner to the Sun cycle of the seasons. That is, during the whole length of the period, the expansion or contracting influence - 2 weeks in the moons case - is taken into account. The Apogee, perigee and nodes periods on the other hand, are seen as periods of time isolated to within 48 hours of the exact point.

A general rule in deciding what moon rhythm to use, is to consider what season you would normally tend to your particular job, summer or winter and then look to the corresponding moon cycles.

The five Moon rhythms do not occur at the same time and often will counteract one another. It is up to you to decide which is the stronger. It is very possible at times they balance each other to a point of equilibrium, in which case you can do as you please.

Phases of the Moon - A general overview

The phases of the moon are significant points in the cycle of the Moon's journey around the Earth. One circuit of the moon indicates a completed cycle of time. This momentum can be used to consciously help any process of development reach fruition

The New Moon is the beginning of the cycle and is the time to picture what it is that is to come into being. Be as clear as possible in regard to the details of the situation or phenomena you are working towards. Explore potentials and do what is necessary to make it occur.

In Plants, the New moon has been shown to be a fairly barren period. Plant matter dries quickly. Seeds can be sown in the right constellation a few days after New Moon onwards. Aerate soil, Mow hay for quick storage, Begin spraying 501.

The First Quarter is the first challenge to this impulse and highlights a time to consolidate the initial exploration that has taken place since the New moon. There can be a feeling that if things are not consolidated they will disperse into many different possibilities.

The Full Moon is the peak of the cycle. The point where the impulse crystallises, but not yet manifest. A stage however, where one can determine what is possible to achieve on this cycle. Major hurdles can also surface. The culmination that takes place, indicates the tasks that need to be carried out over the remainder of the cycle, to bring manifestation.

For plants, before the Full Moon is seen to be an excellent period for sowing seeds.

Following the Full Moon and until the Third Quarter, the decision has been made, and now the details have to be taken care of. The Third Quarter asks for further clarification of the process, so that the last phase up until the New Moon can be used in completion of the tasks. This clarification also allows for reassessments to take place, so that the next cycle, beginning with the New Moon, can begin clearly and be well directed.

After the Full Moon, begin to transplant, prune, spread compost, make compost heaps, Spray 500

Large projects both physical and psychic, can take more than one moon cycle to complete. Each moon cycle can be used to complete definite stages of the whole process. There are thirteen Moon cycles within one Solar cycle. Just as the Solar influences of the Seasons dominate in nature, so does the Solar year override the moon cycles for us. The Moon cycles throughout the year work within this greater framework.

Moon through the Constellations

The plant types are based on the moon moving through the Constellations of the zodiac. The element of the constellation i.e. Earth, Fire, Air or Water: relates that constellation to either the root, the seed, the flower or the leaf.

ELEMENT	Fire	Air	Water	Earth
PLANT	Seed	Flower	Leaf	Root
CONSTELL	Aries	Gemini	Cancer	Taurus
	Leo	Libra	Scorpio	Virgo
	Sagitt.	Aqua.	Pisces	Capricorn

Any work on plants and soil at that time can enhance that basic quality. Plants are categorised by which part of the plant is the desired end product.

Leaf plants -lettuce, cabbage, cauliflower, and ornamental foliage plants.

Root plants - carrots, turnips, potatoes and kumara.

Seed plants -are anything that fruits. Corn, tomatoes, apples

Flower plants - Roses, azaleas, perennials and annuals.

If the plant is predominantly a leaf plant but you wish to save the seed, attend to it during seed days.

Sow , weed,, and harvest all in the appropriate period..

Ascending and Descending

As the Moon's path is 5 degrees inclined to the ecliptic (or Sun's path) , it will spend two weeks moving in an upward direction with regards to the ecliptic and two weeks moving in a downward direction. The Descending phase of the moons cycle occurs when its path across the sky moves lower every day. The ascending period is the days when the moon's path rises higher in the sky. This rhythm causes contraction in the case of the Descending period and expansion when the Moon is Ascending. Tasks such as pruning, taking cuttings and wrenching are best done during this the transplanting period. Apply 500 then also. It helps to activate soil life. The ascending period is best used for sowing seeds, aerating the soil, grafting and applying preparation 501 to the leaf areas of the plant.

In the northern hemisphere this rhythm will naturally be reversed from how we experience it in the South.

The Nodes

When the Moon is passing the Nodes it appears to have a detrimental affect on seeds sown at that period. It is generally thought that this is not a good time for any gardening activities.

Apogee and Perigee

Both of these are seen as extremes that interfere with plant growth. The Apogee days tend to be clear and sunny. However any seeds sown on these days tend to rush to seed before they should. The perigee days on the other hand tend to be still and cloudy, if not oppressive. Seeds sown at this period find it difficult to really find their place in the air and often die of fungal or insect attack. Again some care is needed on these days. A marked increase in the occurrence of fungal attacks has been noticed to occur at this period.

For a thorough introduction to Planting by the Moon, Maria Thun's publications are recommended.

Caution

Remember the Moon is only one part of one layer of the game. The workings of the Ethers, elements, chemicals and soil condition are all to be taken into account when working with the constellation's influences. Steiner pointed out the moon forces are strongest when it is wet, while Saturn works strongly in hot dry conditions. Glass houses show how the atmosphere of an environment can totally overpower the workings of the constellations. This is a multi-layered reality and the constellations are the most distant part of it. refer Earths Layer Cake

The Animal Kingdom

Steiner outlined the animal kingdom in 12 divisions or Phyla. He allotted these phyla to specific zodiacal constellations.

Through the work of several Geothean scientists a colourful picture of the animal kingdom has been developed. The various Phyla or species are seen as remnants of the various developmental stages Humans have moved through.

From this study a key to understanding the place and role of each animal or pest can emerge. E. Kolisko in the work entitled "The

Twelve Groups of Animals" (21) has outlined this view very concisely. Other writers, including Popelbaum and K.Konig have also covered the same topic. (refer Bibliography)

The animal phyla can be divided as follows:

INVERTEBRATES

Protozoa	Single celled	Cancer
Coelenterata	Hydra polyps & medusae	Gemini
Echinodermata	Starfish & sea urchins	Taurus
Tunicata	Sea Squirts and salpae	Aries
Mollusca	Shellfish and snails	Pisces
Vermes	Worms - segmented bodies	Aquarius
Arthropoda	Insects, spiders, crabs	Capricorn

VERTEBRATES

Pisces	Fish	Sagittarius
Amphibia	Frogs	Scorpio
Reptilia	Snakes, lizards	Libra
Birds	Song, prey and running	Virgo
Mammalia	Ruminants, Prey, Rodents	Leo

The Animal Kingdom and the Zodiac.

Initially the animal world can be divided into Invertebrates - soft bodied with no skeleton- and the vertebrates - beings that have a true skeleton. This division suggests that the invertebrates relate to the Moon and the Vertebrates relate to the Sun polarity. As a synopsis to this large and intricate study, E. Kolisko offers a wonderful overview in which he outlines the major function of each phylum and its relationship to the whole. He suggests that there are three steps along the path of development that lead to the human. In each step the same process occurs at a higher level. Within each of the steps, there are four stages

Step 1
Protozoa -   The cellular system is developed
Coelenterata -   The digestive system is developed
Echinodermata - The rhythmic system is developed
Tunicata -   A harmony of all the organs occurs however still at an embryonic stage.

Step 2

Mollusca -	The reproductive system dominates
Vermes -	Digestion process are the dominate process
Insecta -	Respiration is the dominant feature
Pisces -	The blood circulation is developed. Early stages of the heart exist and the harmony of all the organs is at a higher stage

Step 3

Amphibia -	The reproductive system again is highly developed
Reptilia -	The digestive process are again emphasised
Aves (Birds) -	The respiratory system is further developed
Mammalia -	Animals that have a true heart, warm blood.

This outline suggests that the same process takes place at different levels. So in the Coelenterata (polyp), Vermes (worm) and Reptiles (snake), the development of the digestion occurs at different phases of development. Often within each phyla there are subgroups that also emphasis the one sided development of each particular physical system.

The Earth signs:

Taurus -	Echinodermata, Starfish	Rhythmic
Capricorn -	Arthropoda , Insects	Respiration
Virgo -	Aves, Birds	Respiration

The Water signs:

Cancer -	Protozoa, One cell	Cellular system
Pisces -	Mollusca, Shellfish	Reproductive
Scorpio -	Amphibia, Frogs	Reproductive

The Air signs:

Gemini -	Coelenterata . Polyps	Digestion
Aquarius -	Vermes, Worms	Digestion
Libra -	Reptiles , Snakes	Digestion

The Fire Signs:

Aries -	Tunicata, Salpae	Harmony of organs
Sagittarius -	Pisces, Fish	Harmony of organs & blood
Leo -	Mammals , Ruminants	True heart and blood.

The above diagram indicates the correspondences of the systems to the elements indicating that the system development is at least archetypally related by element. The reasoning and phenomena, leading Kolisko to present this correlation, is well worth examining.

Other books on this subject are:

"Man as Symphony of the Creative Word." ..... Rudolf Steiner
"A New Zoology" ..... E. Popelbaum.

Some years ago a homeopath friend used these correspondences as a basis for treatment and found them to be very effective.

The Human Kingdom

The physical body is often divided up into twelve

Aries - Head
Taurus - Neck
Gemini - Shoulders
Cancer - the Chest
Leo - Heart region
Virgo - Intestines
Libra - Renal area
Scorpio - Genitals
Sagittarius Up. legs
Capricorn - Knees
Aquarius -Low legs
Pisces - Feet

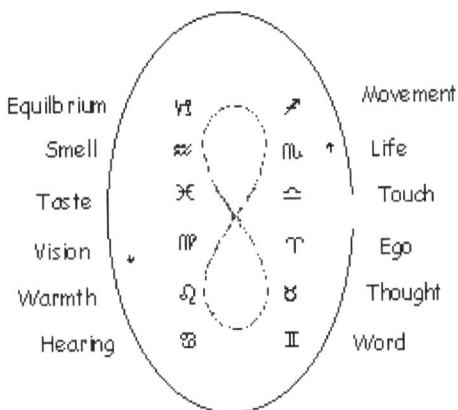

The Twelve Senses

RS developed work on the Twelve senses. From the little I have studied of this work it seems most appropriate to approach it using the lemniscate form.

Using the lemniscate in relation to the 12 fold process is outlined in "Equinoctial and Seasonal Zodiacs"

Seen in this context the natural flow of the zodiac is maintained, while the planetary polarities from the Astrological theorem are also evident.

RS has two ways of working with this model. One starts at Libra and goes in a continuous pattern around as if it were a circle, breaking it into three different zones or groups of senses. Libra to Capricorn are the physical senses, Aquarius to Leo are the Souls senses, while Cancer to Aries are the Energetic senses.

The second uses the development of the senses as a polarity relationship between the signs as shown in the astrological model. eg, Aries \ Libra, Taurus \ Scorpio, Sagittarius \ Gemini.

Yet another way RS used the signs is in his work on the Twelve philosophic standpoints.

Conclusion

From all the above it is apparent Rudolf Steiner was an excellent Astrologer and that Anthroposophy sits squarely and solidly on the exact same archetypal base as Astrology. I trust there is enough presented here for you to see the patterns and use of the Astrological theorem as applied to Biodynamics. The real use of this process comes from how it is used to address specific Biodynamic questions.

In more recent years I have enlarged on the Spirit's role in my essay 'The Spirit in Biodynamics', available from the Glenopathy website.

In the earlier sections we saw how a sense of overview can be created from the Astrological model. From looking at known data, we have seen how the Astrological model describes an ordered structure for **'what is'**. Please refer to the BD Decoded chart at the www.garudabd.org for a fuller picture of how all the Biodynamic information fits on one spiralling diagram.

WHICH CONSTELLATION?

B.D. uses the zodiac for determining planting and harvesting schedules as well as the time of preparation application. It is generally used very effectively. However, it does leave some questions unanswered, especially when deciding on pest and fungus spraying or peppering times.

Most of the work on planting by the moon has been carried out by five or six researchers with the most prominent and influential being Maria Thun. Over the fifty years that this research has been taking place the statistical findings of these researchers have stayed consistent with Astrological theory. It does vary from astrological folklore though, which is often a collection of individual impressions rather than scientific analysis.

There has been one short fall in this BD research though, and this is its ability to define the influence of individual constellations with in an elemental subgroup. In Astrological writings each 'Sign' of the zodiac has very individualistic characteristics, and so too, this must be with the constellations. The Astrological model gives some suggestions as to the quality of the individual constellations.

To gain a cross reference, one needs at least two parameters. BD uses only one reference for the zodiac, the elements. From the model it is apparent that the Zodiac constellations are made up of several

Level 3 -The Modes
The Part of the Plant

The Type of Plant	Nerve Sense Root Fixed	Rhythmic Leaf Mutable	Metabolic Flower \Fruit Cardinal
Fire- Seed	Leo	Sagittarius	Aries
Air - Flower	Aquarius	Gemini	Libra
Water-Leaf	Scorpio	Pisces	Cancer
Earth-Root	Taurus	Virgo	Capricorn

Level 4 - The Elements

components. We have looked at the elements and seen they relate to the plant types at an Etheric level and work with all that level 4 represents. Fire constellations influence the seed and fruit processes, Earth constellations relate to the root, etc.

The Zodiac is also made up of level 3 , the Modes, physical parameters and level 5, the planets providing Astral relationships.

Level 3 in particular divides the plant up into the physical systems of Nerve/Sense, Rhythmic, and Limb/Metabolic systems. This would suggest that any work done according to the constellations will not only effect the plant type, etherically, it will also effect the specific physical system of the plant, indicated by the mode - Cardinal, Mutable Fixed. When one constellation is favoured by accident or willfully, it would tend to imbalance or enhance one physical system of the plant. It could be suggested that a particular crop could be worked or sprayed, in all the modes of a elemental grouping. This would evenly strengthen the physical organism as well as the Etheric sphere indicated by the element. This could best be done within one moon cycle, however it would no doubt be just as effective over an extended period of time.

In fungal and insect attack it is possible to determine the constellation that is most likely to be effective, as one part of the puzzle.

Look to see what element governs the Plant and what mode governs the specific area of the plant that is being attacked.

A cross reference is established that indicates the specific constellation for preparing and administering your solution.

For example; In the case of White Butterfly. It is a caterpillar that effects the leaves of brassica plants. The brassica are predominantly Leaf vegetables and therefore associated with the Water constellations. The part of the plant most affected by the grub is generally the leaf. This is the rhythmic system of the plant and corresponds to the Mutable constellations. Together Water / Mutable is Pisces. Any remedy applied for this problem should therefore be applied when the Moon is in the constellation of Pisces.

THE QUESTION OF THE ZODIACS

The tropical signs or the sidereal constellations?

All stars are focuses of spirit. They all consume Hydrogen and emit forces and generate fields constantly. These forces are the primary formative forces of our Universe. Our Sun is our local World spirit. It therefore is and indicates our source of life. When it is placed in a birth chart it becomes an image of the internalised spirit (Level 1 in BD Decoded) of the individual as apart from all the other stars of the Galaxy which stay as the Cosmic or collective sphere of spirit (Level 6).

So what is the difference between the signs of the zodiac and the constellations of the zodiac?

Tropical Zodiac - The Signs

Firstly it may help if we gain an understanding of the nature of these zodiacs. The zodiac commonly talked about and seen in modern western Astrological publications is called the Tropical Zodiac. This is based on a **30 degree division of the SUN'S path** , along the Ecliptic. Being a Sun based phenomena it is therefore a solar system based reality. If we wished to become finicky we could say it is placed between the Earth and Mars which is where we imagine the Sun to be rotating around the Earth.

The division of this zodiac begins each year when **the Sun crosses the vernal equinox point of the Northern hemisphere.** This occurs around the 21st of March each year. This point becomes the first degree of the sign of Aries. The ecliptic is then divided into twelve equal divisions of 30 degrees each. These twelve signs have the same

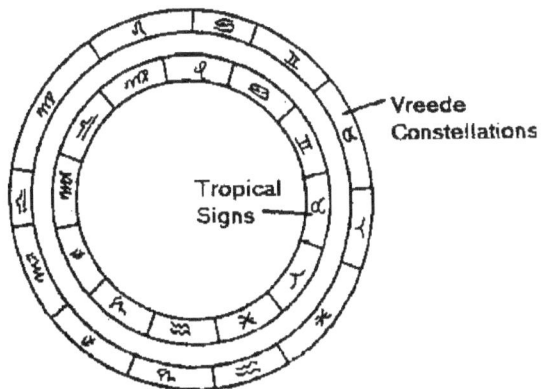

Vreede
Constellations

Tropical
Signs

The Two Zodiacs

names as those given to the constellations. These are named Aries , Taurus, Gemini etc. The important point here is that this 'signal' tropical zodiac is set in relation to the Suns yearly movement, as seen from the Earth, in relation to the seasons of the northern hemisphere.

So this is again an apparent reality. Based more in archetypal law than on any Astronomical realities.

The only astronomical fact we can find is that the Tropical Zodiac is based on the apparent movement of the Sun past a Spring equinox point. In reality this is the Earth moving past the equinox with the Sun.

So with this zodiac we are using the Sun as the focus, not the constellations. The divisions of the zodiac are a division of its path into twelve archetypal divisions.

The Sidereal Zodiac - The Constellations.

The sidereal zodiac is an astronomical reality and is based on the visible stars of the heavens. The constellations used in this zodiac, are those the Sun and the planets move in front of, during their revolutions.

As we are existing on the surface of the Earth and therefore the subjective centres of our universe, both these Zodiacs are Earth centred in orientation. Therefore it appears the Sun moves around the Earth and not the reverse. The stars within the constellations only have a slight movement of their own, however this is minimal and will not be considered in this discussion.

There are at least three different divisions of the sidereal Zodiac. The origin and division of each is well described by Robert Powell in his book "The Sidereal Zodiac". There are two based on the 30 degree division of the 360 degree arc, along which the zodiac is placed, with the difference occurring according to their beginning point.

The other is the unequal division of the heavens used in the Planting and Sowing Calendar of the BD organisations. These constellation divisions are astronomically correct and are therefore of varying sizes. Therefore the passages of the planets through the different constellations take varying amounts of time. This unequal division of the Sidereal zodiac is based on the work of Elizabeth Vreede in the 1920s and more recently on practical agricultural experiences of Maria Thun and Agnes Fyfe. The divisions of this Zodiac remain constant with the Sun and the planets moving through it.

The Vernal equinox of the Sun has no influence on the beginning or ordering of the constellations. It is the vernal equinox which moves in front of the constellations and has been used for centuries - the great marker of the ages and human evolution. The vernal equinox, according to the unequal division of the constellations, occurs at 6 degrees in the constellation of Pisces at the present period. Due to the phenomena of the precession of the equinoxes, the vernal point is slowly moving backwards, towards the constellation of Aquarius at the rate of 1' every 72 years. Using this system the Age of Aquarius is therefore not due to start for another 432 years. Other divisions of the constellations put the Vernal Point at 4 degrees Pisces and hence the Age of Aquarius is slightly closer.

The connotation of all this

We must appreciate the massive Astronomical differences between the Sidereal and Tropical zodiacs. The sidereal is based on a twelve fold division of the Galaxy - the constellations.

While the tropical - the signs - is a twelvefold division of the Solar system. The beginning of this division is based on a seasonal consideration of the northern hemisphere of the Earth.

Hence the Constellations are a Cosmic Spirit based phenomena while the Signs are a World Spirit based phenomena, being based on the Sun and the Earth. Essentially occurring within the Astral sphere.

Approximately 2000 years ago (215 AD>) around the time of the incarnation of the Christ the tropical and the sidereal zodiacs, both started at the same point in the sky at 0 degree of the constellation of Aries.

Alexander the Great in 331BC introduced to the Greeks the idea of the Ecliptic based Zodiac he found in Babylon. Hipparchus promoted its use in the 2nd century BC and was ridiculed. Ptolemy (200AD) was the one to officially institute the Tropical Zodiac as a valid Astrological model. He luckily lived at the time the vernal equinox was exactly at 0' Aries. Generally prior to this period, the Sidereal Zodiac was used. The early astronomers of the Chaeldean and Babylonian cultures developed their knowledge around the constellations. The stars Alderbaren and Antares were used to mark the middle of the constellations of Taurus and Scorpio respectively and from here a 30 degree division was made.

Christ appeared in the middle of this period. Rudolf Steiner illustrates clearly that the appearance of Christ marks an intensification and central point for humans developing as a energeticly independent ego conscious beings. As this continues we can expect to evolve to the level of creative angelic beings at the end of this present period of Earth evolution. (see BD Decoded) Prior to the Christ's initiation, man relied strongly on the guidance and assistance of the various levels of energetic beings. This can be seen in the Myths and legends of early civilisations of our planet. In the Indian scriptures the energetic beings came and went from the Earth with ease. Their presence was part of the common reality to all the inhabitants at the time. As Mankind began to incarnate onto the Earth more firmly, he began to loose contact with the Gods and by the Egyptian period he was trying to develop methods of preserving this knowledge for later Mankind. The Egyptian- Chaldean culture marks the last period of direct energetic communication.

The battle of Troy (1500BC) also contains many images suggesting a time the Humans were awakening to an individual life separate from the Gods. After Troy fell the Greek gods retreated more and more away from the humans. It can be surmised that the Humans had finally had enough and expelled the Gods. A study of the immediate history following Troy reveals there were no winners in Troy. Classical Greece self destructed into chaos, immediately following Troy, and when it came together again 100 years later, we see the rise of Hellenistic Greece with the development of rationalism and human control. In subsequent cultures the study of Man and the physical Earth becomes more important.

RS points out the Christ was a being of the "Sun sphere" and that the development of Man as a energeticly independently connected being was an activity connected intimately to the Christ as messenger from this sphere. The Sun, being the life producing star of our system is seen as an image of pure internalised spirit. On the chart, this is Level 1, the point where all existence merges into the spirit.

At the crucifixion, the Christ connected with the Earth in such a manner as to be considered the spirit of the Earth. Here the macrocosmic internalised spirit sphere connects with the Earth. In turn so individual humans can incarnate the spirit consciously, thus having a direct personal connection with "All that is".

The Christ as the messenger of this sphere encourages us to find a direct relationship with the Godhead, a relationship that goes beyond all class, race or financial barriers. We are all worthy of the spirit. Indeed the incarnation of the spirit will lead to the awakening of our individuality and objective consciousness, with positive and negative influences, depending on the morality of the individual.

It is interesting to note that the Sun oriented tropical zodiac began to appear around this time. As Humans have continued on their path of individuality it has increased in importance especially for Western cultures, (to the point where very few people, even in 'educated circles' know the difference) where the Ego incarnation has been the greatest.

The development of the Ego in the Human is a quality or energetic member that is unique to the Human and something they do not share with the plants and animals . The plants and animals have an Ego impulse, however this works on a group rather than on an individual level. It does not incarnate but remains outside the physical form of these kingdoms. This leaves the plants and animals in a direct energetic connection with the environment and their larger environment the Cosmos. Their Ego influence continues to stream to them from the Fixed Stars directly. Humans, by developing their Ego consciousness, loosens their connection to the immediate cosmic environment. By internalising the Fixed Star influences we have the potential to become liberated from many of the direct impacts of the Fixed Stars. Thus using the constellations as a background for the birth chart would connect us more to our pre-Egoic, less individualised state of consciousness.

On a unconscious physical level the constellations still have a strong impact on Humans. Current Indian astrologers still use them to good effect. On the objective conscious level though, we do now have a greater potential to make individual decisions over these influences. On a gross level the manifestation of genetic engineering highlights how the archetypal forces sourced from the Fixed Stars can now be consciously manipulated to our ends on Earth.

Cosmic rhythms and movements remain in an archetypal form within man. If our freedom of will is not recognised consciously, the cosmic events still have a strong influence on Human destiny. Elizabeth Vreede points out in her Astronomical letters that the Human journey through the energetic world is under the influences of the constellations until one incarnates on Earth when the Human again comes under the influence of the Tropical Zodiac.

The Earth is the only sphere where Humans can work through past karma and develop further on the present path of energetic growth. During the journey between death and rebirth the Human is free of the Earth and its destiny and once again is a cosmic being. This allows one to come in direct contact with the energetic hierarchies and the cosmos. This places the Human in clear connection with the Sidereal zodiac during this period. Since the Christ's appearance on Earth, the destiny of Humans has been more strongly connected to the Sun and the energetic beings of that sphere. The Human, however, must now ascend to them, rather than them to us, as the Christ did.

So hence when dealing with **Human destiny in the course of our existence on the Earth**, it is appropriate during this period of "Ego consciousness", to use the Sun oriented tropical zodiac. Not only are the Signs Sun based, they are based also on the subjective reality as the seasons of the northern hemisphere. A very Earth orientating reference. Strengthening the fact the Signs relate to life on Earth. **Hence the Tropical signs indicate the incarnated spirits life on Earth.**

The Human is striving for the point of guiding our own destiny and is therefore freeing ourselves of the direct cosmic impulses. It has been my experience, during 20 years of Astrological consulting, that the tropical zodiac (and a further division, the houses) indicate clearly the karmic events and growth stages an individual must experience while on the Earth. The constellational chart indicates a picture of the macrocosmic collective spirit being used to assess the more unconscious destiny of the individual.

Using this thesis then, the constellations are best employed when working on the spheres of life strongly connected with the cosmos, without any suggestion of free will. We have seen how Plants are embedded in the body of the cosmos. They embody an Etheric body with their Astral and Ego aspects still external. As such the planets and fixed stars act directly as these 'organs' of the plant kingdom, so their movements affect plants in a direct manner.

The Human, on the other hand, is in the process of internalising the Ego, and in doing so can exercise a degree of liberation from the exact movements of the cosmos. In so doing we have the opportunity to become an individualised microcosmic double of the macrocosmic universe.

Further questions need to be asked - What about the Southern Hemisphere? Will the signs activity be different? Should we use a Zodiac beginning with Libra and ending with Virgo in the Southern hemisphere? These are big questions. One justification for retaining the Northern hemisphere equinoctial beginning for the zodiac is that the Earth is polarised magnetically. We do have a magnetic north focus and hence a universal beginning point. The zodiac however is so arbitrary and so Earth bound it does stand to reason that we are able to develop a southern hemisphere zodiac if we wish. It is only one small step along an already extremely abstract road. Many discussions have been had amongst southern Astrologers with a general opinion being current that Libra does act as a spring , assertive sign with many 'Aries" qualities, here in the south.

See 'Signs and Constellations'

http://garudabd.org/wp-content/uploads/Signs-and-Constellatons.pdf

EQUINOCTIAL & SEASONAL ZODIACS

To add to the debate of which zodiac to use, we are confronted in RSs work with the question of the zodiacs. He presents us with several other zodiacal beginnings as well as zodiacs going in the opposite direction. In the work on the evolution of the animal kingdom, we meet Cancer as the beginning, while in Hauschkas work on minerals he gives a Leo start to the zodiac. In one other work associated with human social development I have noticed a Virgo starting zodiac, while the twelve senses work uses a Libra starting zodiac. In all these zodiacs the description of their application is through the zodiac in the opposite direction to what we are use to.

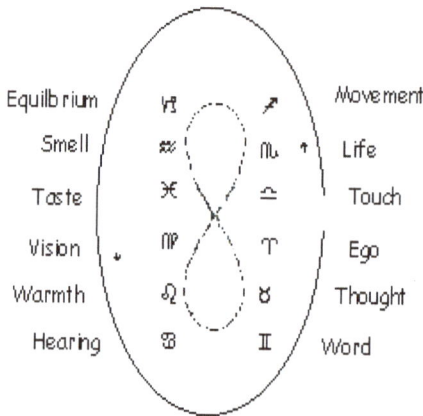

The Twelve Senses

The Archetypal and Manifest Zodiacs

The Astrological model identifies the Cancer to Leo zodiac as possibly having some special significance. It presents us with a different structure to the zodiac, than the 'traditional' anti clockwise Aries through Pisces zodiac, used by most of Astrology. Initially the pattern looks chaotic even meaningless however as we look into it more there is an order that arises. As described in Astrological Science -12 fold, (pg 52) there is a lot of order in this model and we can see all the laws from the earlier levels of the Astrological spiral are carried through in this patterning.

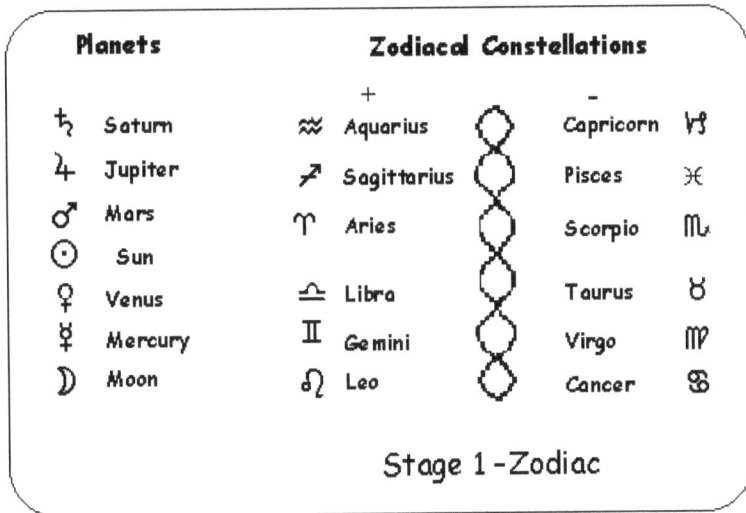

Planets		Zodiacal Constellations			
		+		−	
♄	Saturn	♒	Aquarius	Capricorn	♑
♃	Jupiter	♐	Sagittarius	Pisces	♓
♂	Mars	♈	Aries	Scorpio	♏
☉	Sun				
♀	Venus	♎	Libra	Taurus	♉
☿	Mercury	♊	Gemini	Virgo	♍
☽	Moon	♌	Leo	Cancer	♋

Stage 1 - Zodiac

Being of the Astrological model we can expect it to be archetypal in nature and therefore to describe founding principles of whatever it is applied to.. From this model we can see Cancer and Leo stand out as the only signs ruled by a planet each. We can note however that the Sun and Moon form a creative polarity with each other at level two of the Astro. Model. Following Koliskos lead given to us in "The Twelve groups of Animals" and that Mother Cancer can easily be seen as the starting point of life we can track the development of this zodiac by moving in a clockwise direction going from Cancer > Gemini > Taurus.....through to Leo. If we follow this line of order starting at Cancer we see the double helix comes into form. This mimics a DNA like series of twists.

In " The Twelve Groups of Animals" Kolisko interprets the animal kingdoms evolution from Protozoa to Mammals through this twelvefold process starting with Cancer running backwards through to Leo. This outlines a fundamental manifestation of the archetypal law. We are not talking about the patterning of hair on a dog or feathers on a bird, which is more according to fibonacci number series, but a more fundamental patterning of the animal evolutionary archetype.

The next question is, how does this zodiacal patterning relate to the circular zodiac we are use too? We are given a hint about this process when we see Steiner's "Twelve Senses" zodiacal patterning. This uses a patterning of the zodiac based upon the lemniscate. This zodiac though starts with Libra and not Cancer. We can however note the principle that zodiacs can be organised as a lemniscate.

In the familiar circular zodiac we have an arrangement of signs around the circle in a positive and negative order. Aries +. Taurus -, Gemini + etc. It is also known that all living things and many minerals carry an electrically charged polarised form. This principle is carried right down to human life, needing a male and female sexual polarity to come into form. So it can be stated that polarisation is the natural state of manifest forms. In the stage one zodiac this polarisation is 'external' . We are defining the zodiac by the dual rulership of the planets. Each planet of the model rules two signs of the zodiac and so on. So if we are to head towards the polarised manifest circular zodiac, as it appears in the sky, then this polarisation must be internalised.

The 'stage one' diagram still has externalised polarity, compared the stage 3. That is, the positive planets are placed on one side, the negative planets are on the other, mimicking the polarity law. If this pattern was placed directly on a circle we would have the positives on one side and the negatives on the other. Interestingly this is what we find also when we put the periodic table on a circular format (see pg 191). Cations on one side and Anions on the other. Just an observation.

To take the next step we can see a lemniscate form can be made in this zodiac through the 'unwinding' of the mutable signs. From

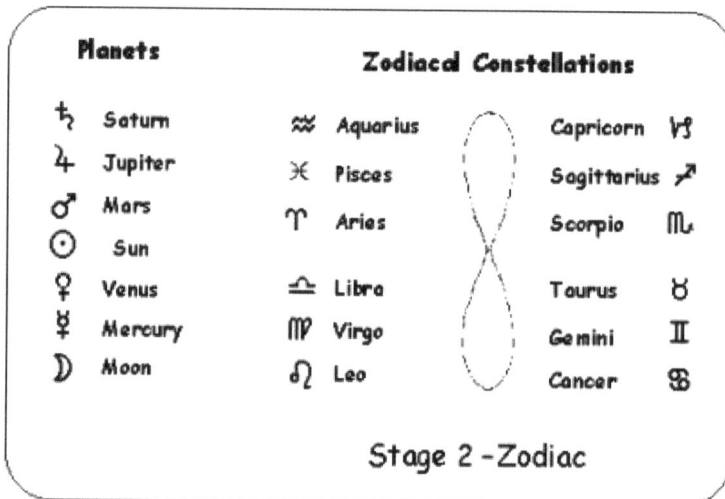

Planets		Zodiacal Constellations		
♄	Saturn	♒ Aquarius	Capricorn	♑
♃	Jupiter	♓ Pisces	Sagittarius	♐
♂	Mars	♈ Aries	Scorpio	♏
☉	Sun			
♀	Venus	♎ Libra	Taurus	♉
☿	Mercury	♍ Virgo	Gemini	♊
☽	Moon	♌ Leo	Cancer	♋

Stage 2 –Zodiac

Stage one has this same internal modal order as stage 3, however they are all positive modes or negative modes sitting together. The most moveable part of the Zodiac is the Mutable signs. This is their function, to mediate between the two poles of Cardinal and Fixed. So to unwind this spiral we can first of all unwind the mutables, and in doing so we create an internal polarisation within the four quarters of the lemniscate. The negative Pisces now sits between the positives Aquarius and Aries.

This patterning can be explored further with regard to a twelve fold planetary structure and the Human energetic path. - see the Glenopathy website. In this example and the 'twelve senses' we see Human psychology and energetic development described. These are pretty abstract realities and a little removed from the vital basics of manifestation. It could be argued that if these things did not happen the rest of the planet could not care less.

So we can suggest stage two - the lemniscate - indicates a pattern or activity that stands slightly removed from the archetype and yet still behind manifestation. It appears to be an intermediate stage between these two.

The third step of this process is to unwind the bottom half of the lemniscate to create the zodiac we are use to. This zodiac, while often used for describing a 12 fold personality division, is more traditionally

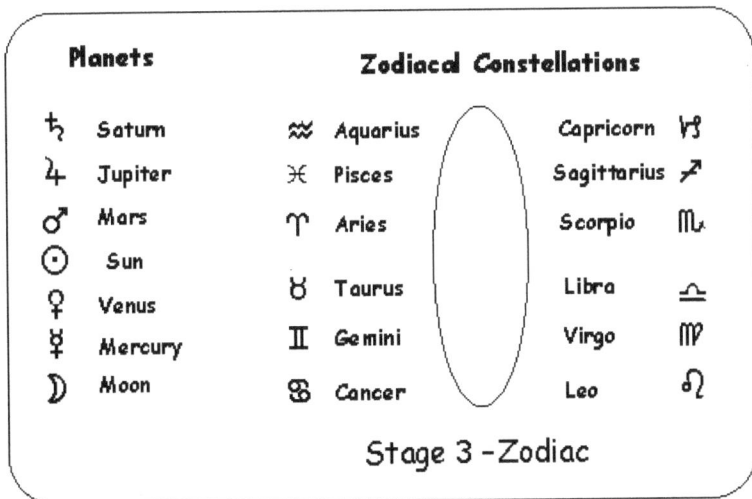

Planets		Zodiacal Constellations		
♄	Saturn	♒ Aquarius	Capricorn	♑
♃	Jupiter	♓ Pisces	Sagittarius	♐
♂	Mars	♈ Aries	Scorpio	♏
☉	Sun		Libra	♎
♀	Venus	♉ Taurus		
☿	Mercury	♊ Gemini	Virgo	♍
☽	Moon	♋ Cancer	Leo	♌

Stage 3 – Zodiac

used to describe the physical body. Aries - head, Taurus -Throat, Capricorn-knees etc.

This now brings us to another big question. What is the relationship between the Cancer >>>> Leo zodiac and the Aries>>>> Pisces zodiac?

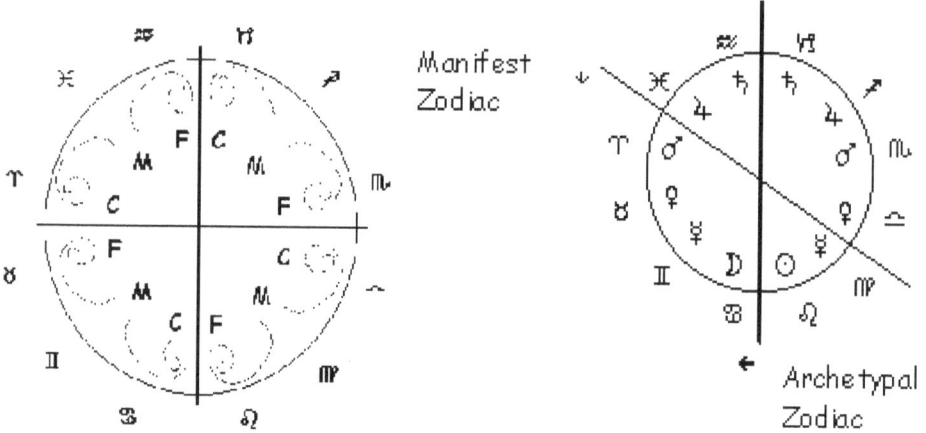

Manifest Zodiac

Archetypal Zodiac

The zodiac division we take for granted is the Aries >> Pisces zodiac, which we saw in the last chapter is often related to the physical manifestation of the seasons. Even more specifically of the northern hemisphere. So before folks in the Nth hemisphere get too carried away with drawing associations between seasonal activity and archetypal zodiacal phenomena, please remember the north's experience is not ours in the south.

To gain a clearer picture of these two zodiacs we can put the traditional planetary rulers on the 'finished' stage 3 zodiac. Look at where the Aries zodiac starts from a planetary point of view. Is it not a little strange it starts at a Mars Jupiter cross point rather than the Sun Moon mid point or at least the Saturn mid point or the Venus Mars mid point. The zodiac according to the modes presented earlier also supports the notion put forward by this 'planetary' zodiac.

THESE DIAGRAMS INDICATE THE ARIES >>>>>> PISCES ZODIAC IS NOT ORIENTATED BY THE ARCHETYPAL MODAL, PLANETARY OR ZODIACAL LAW BUT BY THE PHYSICAL SEASONS and hence highlights further its relationship to matter and a physical earth life than the archetype of spirit. It is a manifestation zodiac.

The only apparent anomaly of this planetary and zodiacal structure is that Cancer and Leo are ruled by the Moon and Sun. The Sun and the Moon make up the primary polarity of life (level 2). From our view they are the same size and are far larger than any other of the heavenly bodies. They are set apart in all mythologies, from the other planets as Mother and Father, Rangi & Papa, Yin and Yang, beginning and end. Cancer in Astrology is the mother and the source of life, while Leo is Lion the King of the Jungle, the peak of one personal creative expression. This Cancer >>> Leo zodiac structure again places the Sun and Moon as prominent points of the natural cycle by being the beginning and end. Worthy of individual rulership of their signs and imaging the primary polarity we found as the source of life at level 2.

Quite unconsciously we do use this zodiac in modern Astrology. The two things that do travel backwards in traditional Astrology are the Precession of the Zodiac and the movement of the north node.

Precession of the Equinoxes

The Precession of the Equinoxes, is the name given to the backwards movement, through the zodiac, of the position of the Sun at the Spring Equinox, in the Northern Hemisphere. This phenomena is the only major use of the constellations remaining in most Western Astrology. It is from this spring equinox point that the signs begin and are marked off in equal divisions. They are however not a division of the constellations, they are a equal 12 fold division of the ecliptic; the Suns apparent path around the Earth, as outlined earlier.

The position of the Spring Equinox moves 'backwards' through the constellations at the rate of 1 degree every 72 years. It takes approximately 2160 years to move through one equally divided constellation and 25,920 years to go once around the zodiac. At present we are placed between 4 and 6 degrees of the constellation of Pisces. - depending on the constellational division one uses. It is this cycle from which we arrive at the concept of the age of Pisces and Aquarius. This equinoctial cycle has been used by many scholars to correctly indicate the developmental history of humanity. It not only indicates the cultural development, it also provides the structure for understanding the energetic evolutionary development of humanity as well.

At the earliest we will not have the equinox in Aquarius for another 300 years. Allowing for the orb effect of 3 degrees we are still over 100 years away from moving into orb of Aquarius. So where is all the Aquarian energy of our present time coming from. The quick answer is the Saturn, Uranus Neptune conjunctions from 1987 - 1997 and the Uranus and Neptune transits through Aquarius from around 1995 till 2005. We are also in the orb period of Aquarius so its early influences will be effecting us. Another reason is that our evolutionary process is speeding up due to the effects of the pollution of our atmosphere. We, humanity along with all other species, are 'going to seed' early and in doing so our future evolution is being sucked towards us. Thus the Aquarian influences of the future are coming earlier and stronger than they may have if industrialisation was handled in another manner. I suggest that at the present rate of decline of our atmosphere humanity will have had the chance to evolve in consciousness at least, to the end of the 6th post atlantian period by 2012 and to the end of the 7th post atlantian period by 2025. By this time I suggest the free oxygen of our atmosphere may be all but consumed by pollution and that natural human life on Earth may well be impossible from that point. I say our consciousness can evolve as our bodies may be much slower to adjust. Which is why they will go the way the frogs presently are. 4000 years of consciousness evolution within 25 years is a lot of culture shock. Rudolf Steiner has outlined the Cultural Epochs as follows

Constellation   Cultural Epoch

Leo	The last Atlantian epoch		
Cancer	Indian - Beginning of the Post Atlantian periods		
Gemini	Persian		
Taurus	Egyptian		
Aries	Roman		
Pisces	European	5th post atlantian	
Aquarius	Russian	6th	"
Capricorn	American	7th	"
Sagittarius	Beginning of the next cycle of seven civilisations.		

Constellations and Ages

So here in the Cancer zodiac we have an image of the archetypal energetic and cultural development of humanity. As this is the main use of this Zodiac in Western Astrology, I choose to call it the Equinoctial Zodiac as opposed to the Seasonal zodiac we have starting at Aries.

It should be noted our Solar system takes 220 million years to go once around the Galaxy. So this Equinoctial cycle is not based on a Galactic rotation. It is understood that our solar system and seven other Stars are circling around the star Alcyone which is placed in the Pleaides. This cycle takes 25,920 years.

This Precession of the Equinoxes is still a northern hemisphere phenomena. For the Southern Hemisphere we are in the opposite constellation. Our Spring Equinox is in fact in Virgo at present. In the future we are moving towards a Leo epoch. This indeed presents us with a problem and possibilities. Are we in the southern hemisphere in a different cycle of evolution?.

When looking at any particular age though it is interesting to note that Pisces exists within a 'cross' relationship with the signs Virgo, Gemini and Sagittarius. Whenever one of these signs are emphasised all will become active. In the Pisces age we have the Christ as the avatar of this period. He preached open hearted compassion for all beings (Pisces) and a direct personal relationship with God (Gemini). What we have seen develop during this age is the Catholic Christian church is the promotion of chastity, purity, and sin (Virgo) through a dogmatic theology (Sagittarius). So what we have is an impulse of the age with its counterpoint manifesting simultaneously. More examples could be explored.

In the coming Aquarian age, we all have the potential to personalise the message and gift of the Christ. We can all function as a messenger of God. The focus now is upon the Aquarian qualities of acknowledging the God within each of us. The group becomes a collective of 'gods' with the leader of the time arising out of the needs of the time. The challenges will arise from Leo - the desire of one individual to be the supreme 'god', Taurus - the challenge of letting go of what we believe to be ours and to share it with the collective, Scorpio - To trust our fellows with our nuggets of gold.

In the Southern Hemisphere we could look for how Leo is in fact our age impulse. According to the Equinoctial Zodiac this would place us at the end of the cycle. Leo becomes the constellation of full conscious creative awareness. It is in Leo we become our truly creative selves. While it indicates 'the King', it is the beneficent, magnanimous leader who shares the bountiful plenty of our own creativity with all about us. In this scenario Aquarius becomes the 'anti hero'. The dogmatic, set traditional beliefs of the society which becomes the limitation to the creative expression of the individual. So are we in the south following more individual paths, while those in the north are moving more strongly towards a community of individuals?

All of life will give indications of which impulse is appropriate. All we need do is look and see as time moves on. At present the development of global communication, silicon chip technology, the rising people power, are all Aquarian impulses. Likewise with the development of these Aquarian activities there is a dramatic release in the potential for the individual expression, in the forms of more avenues of personal expression through writing, music, art etc. The whole world is now your stage. You have free access. All you have to do is be heard. Both Leo and Aquarius are active here. The fact you are reading this is a testament to this very point.

Again the fact the Earth is polarised to the North pole, may well hold the deciding vote in which impulse is dominate in Human evolution. What do you think?

This brings up for consideration what the connotation of an Earth magnetic pole shift, in 2012 will actually mean. If indeed this does happen. If the poles shift, then this orientation of the equinoxes to the northern hemisphere, due to the north pole dominance of the Earth, will have to be reconsidered. If the south pole now, comes to dominance, then we will need to use Libra rather than Aries as the spring beginning of the zodiac, and the Leo rather than Aquarius as our equinoctial age constellation. All interesting and I guess we will have to wait for the magnetic pole shift and see what happens.

The Moons Nodes

The other major Astrological influence using this 'backward' motion is the Moons Node. These are Astronomical points in space rather than anything 'real'. The Moons nodes are where the Suns path and the Moons path cross. This point , as seen from the Earth moves backwards on a cycle of 18.5 years.

It will suffice at this stage to say the nodes are indicators of the "destiny moments" in an individuals life. Either when planets transit the birth and progressednodes or when the node transits sites in the natal and progressed charts. These are opportunities available to us to pick up if we wish. They are not fated. As such they are moments the spirit can impact upon its course of future evolution, as apart from 'karmic events' coming at us to deal with. Again this indicates an orientation to the spirits possible action, not the Astral bodies fate.

The south node can be seen as the past and the north node as the future. The south node is all the talents and skills one brings from the past and now needs to be used as the basis upon which to build the future- the north node. Some Astrologers give a more fatalistic view of this relationship and suggest the south node must be left behind to attain the north node. I do not support this. One builds upon the other. Yes there is a balance achieved in the process, however the future must be an organic growth out of the past using the past for the development of the future.

The conclusion

From all the above and from the results of investigations into using the traditional and octave planets in relationship to the Equinoctial zodiac I suggest, it finds its activity as an indication of the energetic developmental process of the Human. (see "The Planets") . I am left to conclude the Cancer > Leo zodiac outlines another layer of the archetypal energetic impulses coming into life from the Universe. These are essential energetic formative impulses upon which matter and consciousness accumulate. This is one aspect of Geothes "Being".

The Aries \ Pisces zodiac on the other hand is linked to the seasonal cycle of only the northern hemisphere of the Earth. The spring equinox is seen as the beginning of the year, while Aries is seen as the first sign. So we have the signs interpreted in relation to the unfolding and contracting of the seasons in the Northern Hemisphere.

The unfolding of the 'seasonal' zodiac is used in many ways to describe the path of growth on Earth of any process one wishes to apply it too. One basic example follows: The path of the individual through personal discovery, - Aries to Gemini, to tribal awareness - Cancer to Virgo, to relationship awareness - Libra to Sagittarius, to Social awareness - Capricorn to Pisces, to energetic awareness. This 12 fold patterning can be applied to any process and will accurately describe its 12 stages of earthly development.

The Houses

The Houses provide another example. They are a further derivative from this 'earthy' zodiac and are based on the position of eastern horizon in the birth chart as their starting point. Aries rules the first house, Taurus the second and so on. Being another step towards the Earth - its eastern horizon - they outline very real practical manifestations, of where in ones life any planetary influence will appear. This reality is taken further in the practice of Astro Carto Graphy (see later chapter). This method is where the birth chart is adjusted for an individuals travel about the planet. A chart is drawn up as if the person was born at the same time (for the birth place) yet at a different location. This moves the eastern horizon to another place in relation to the planets. Interestingly the same planetary relationships and activity start manifesting in different ways in the new location. This is a very useful system for geomagnetic manifestation, but in this context just adds proof to the claim that these earth based phenomena accurately indicate physical manifesting influences.

The Seasonal zodiac has been the most widely used in both Constellational and Signal Predictive Astrology for thousands of years. When we look at the planetary rulerships of the signs of this seasonal zodiac, on the circle, they do not represent any inherent pattern, as we saw earlier. Indeed, looking at it this way, one must ask why would anyone start a zodiac at Aries? It is a fact though that the physical body is appropriately described by the Aries \ Pisces zodiac and that foetus development does start with the Head and grow towards the feet.

The fact these zodiacs head in opposite directions must have significance. Alice Bailey suggests the different directions relate to the opposing processes of descent into matter - anti clockwise and the rise of the spirit away from the Earth - clockwise.

Again the polarity between Manifest life on Earth and Energetic impulse is apparent in these two zodiacs. We can conclude - The Aries \ Pisces zodiac appears concerned with material life on Earth - the descent into matter, while the other, Cancer \ Leo is concerned with the energetic unfoldment of the individual and its rise to spirit.

Energetic unfoldment is the point of our evolution as described by the Procession of the Equinoxes and mirrored in our evolution of the Human to full Ego incarnation, especially in this age of Pisces. Now we have this potential - Ego\ spirit consciousness - being at the end of the age of Pisces, how are we moving forward into the Age of Aquarius?

Possibly, as conscious individuals motivated by our individual connections to the spirit, which enables us to function as empowered equals for the good of the whole. Here leaders arise out of the need of the time rather than being based on a personal need to control. The people who we need to be in power will be. With the spirit incarnating we have the potential to moderate and even control the wild Astral body impulses. Our bestial natures can hopefully be subdued enabling a greater sense of empathy for our fellow humans.

Centrifugal and Centripetal forces

In a spiralling motion, the creation of the spiral is maintained through an interplay of two sets of forces. One set moving from the centre to the periphery in a an anti-clockwise centrifugal motion, while the second moves from the periphery to the centre in a clockwise centripetal motion.

Individual researchers, like Schuberger, Schweck and Riech have all found the centripetal clockwise spiral is life promoting. The centrifugal anticlockwise motion was destructive to life. Riechs terms were that the clockwise motion created Orgone energy while the anticlockwise motion created Dor energy. Hence we have the building up of life and the destruction of life.

Using what was said earlier, about Centrifugal and centripedal forces, we can add we have the Equinoctial Zodiac moving clockwise, indicating the archetypal 'life' creating process, while the Seasonal Zodiac, moving anticlockwise outlines the materialising 'death' process. To enlarge on this it will be of value for the reader to explore Geothes concept of creative processes. He outlined how each created being has an incarnation phase and an excarnation phase. The incarnation phase was where the necessary forces for life were being drawn together. This phase is essentially invisible. Then comes the manifest phase of form creation. What we see as a life form is essentially a form already moving towards death. The coming into form brings manifestation, yet the creative forces drawn together in the first phase then begin to move towards maturity and dissipation and seeding. Hence Earthly manifestation indicated by the anticlockwise Seasonal Zodiac, is the last part of this creative process, showing our moving towards maturation and death on the physical plane. The opposite is another aspect of our journey back to spirit.

PLANT PREDATORS

Plant Predators come in many forms. There are the various insect predators as well as many fungi, bacteria and virus's. For all of these Biodynamics accepts the same basic axiom as the basis for their existence. Whenever they cause a problem they are always a manifestation of an imbalance in the relationship of the four bodies to each other. The causes of which are manifold and the solutions as varied as the problems we find. Undoubtedly though it will be some form of stress factor which causes the imbalance.

One imagination I use to visualise the existence of an insect attack is to imagine a finely woven cloth of four different colours. When everything is harmonic no light shines through the cloth. As stress factors mount the basic colours begin to separate into their individual groups. It is as if oil and water are separating from one another. As the separation continues large holes appear in the fabric (of space) and it is to these vacuums in space, that the insect or fungi is actively sucked. The solution is not to kill the insect and leave the vacuum open, for another insect to be sucked into, but to close up the hole. This means to alleviate the physical cause of the stress as much as possible and then to use the Biodynamic preparations to further enhance the energetic body recombination.

The biodynamic preparations have shown that they can be used by themselves as a solution. However their effect will only last for a period of a few weeks before the underlying physical imbalance reasserts itself once again. So solving the physical problems as much as possible is the long term answer.

Due to the symbiotic relationship of living organisms of the Earth and their environment, it is possible to look to the organism for indications as to what the energetic imbalance might be. This is where much of the information provided in the Biodynamic literature is very valuable. Firstly by exploring the nature of the problem we find the common elements between the problem and its environment. With regard to insects and the animal kingdom in general, the work of Eugen Kolisko, (21) which has been presented in the 12 fold section, becomes useful. He has outlined the predominate activities present in the various species of the animal kingdom from which some determinations can be made.

Climate and the soils state is naturally a prime consideration, as these show us the ways the four physical elements of Warmth, Light, Water and Earth are working. During the growing season it is the alteration of these elements which generally cause the primary state of stress. Better drainage, more humus in the soil, mulching and even just watering plants is often all that is needed to bring back the balance.

Insects

Insects were one of the earliest group of species to evolve in a different direction to most other animals on the Earth. It is therefore reasonable to see insects in quite a different light to problems caused by the higher animals.

While adult insects generally still maintain a threefold division of their bodies, into head, abdomen and thorax, they do have a significant difference from other animals in that they literally have air flowing through them. Just as we have our circulatory system for our blood which moves through ever branching veins to our extremities. So insects have holes down the side of their bodies which allows air to flow into a similar vein structure through their bodies. To such a degree that insects can be considered 'air beings'. It is as if they are manifest Sylphs. Wherever we see insects, we see the activity of the air body, the Astrality. RS comments on this in the sixth lecture of the Agriculture course where he mentions birds also being intimately entwined with the Astrality. Birds also generally have hollow bones filled with air.

Insects are also quite rare in their ability to form specific relationships with plants. Often a particular insect will have some unique adaptation which will make it the only insect that can pollinate a particular plant or even only eat a particular plant. Often flowers and their insect pollinators will have a similar shape so as to be indistinguishable until they fly away.

So it is within the bounds of possibility to accept the insects at least as a symbiotic partner to the plant kingdom. Plant evolution has been intimately linked to the life of the insects .

If we look at the energetic make up of the Kingdoms of Nature, presented earlier, we can see that the plant kingdom has an internalised Etheric body, while the animal kingdom, which includes the insects are different to the plants in that they have internalised the astrality. Koliskos work has shown that the degree by which an animal has developed true organs is an image of the degree it has incarnated the astrality. Insect organ development is still relatively fundamental however they are definitely on the road.   From this, we can make the general statement, that plants are beings of the water in the same way that insects are beings of the air. In the relationship between plants and insects we see the relationship between the Etheric body and the Astrality.

Remembering that plants have an internalised Etheric body they therefore live within the 'body' of the World Astrality and World Spirit forces. This means that generally the Astrality lives outside the plants, and so the rightful place of the insects is to hover above the plant, drawing their sustenance from the air and that which lives in the air.

So now imagine that when a plant grows through the spring it is a Physical and Etheric organism, carried on the minerals and water from the soil. It does this all the time pushing back the World Astrality as it grows. Once the summer begins and the water begins to diminish from the environment and the plant, we see the Etheric body beginning to withdraw into the Earth again. From earlier chapters you should by now appreciate that the Etheric and the Astrality are like marriage partners. That while they are of opposite quality, they form a creative polarity which allows both a creative tension from which to become active and 'alive'. In life wherever the Etheric body goes so the Astrality follows. Indeed the Ethericx needs the Astraals activity ti keep it moving and 'in Life'. It can be considered that they are inseparable in life.   So as the Etheric withdraws during a dry period so in comes the Astrality into the physical body of the plant.   Instead of having a physical /etheric matrix of substance, we now have a physical / astral substance. It is this astralised substance which the insects can now move in on and consume.

This suggests that the easy answer to pest control is through the revitalisation of the Etheric activity. The etheric activity returns and the astrality is expelled and the insects have to leave. In the beginning of my studies I worked on this premise and for a small number of insects this solution works. For other insects and plants though, it does not, as there are many things which cause the Etheric to flow or not, in the various plants and many different pests which can attack a plant at various stages of their growth. This brings us back to the need to find the specific nature of both the plant and the insect and for these prime conditions to be reinstated.

One example is my Etherics 7 pest control. It works well on slugs snails and some grubs. By looking into Koliskos work and others we can appreciate that the four stages of the insect, egg, grub, pupa and adult correspond to the Earth , Water, Air and Fire elemental stages. Hence the grub , of which the snail could be considered, is indeed a very etheric being, and hence a stimulation of that energy will counteract that of the being of that energy. However over the years I have noted that it will only repel those insects on particular species of plants and not on others. One cool spring my basil plants were being eaten by slugs while my lettuce and brassica plants where not. Both had been sprayed similarly. I realised that basil likes warmth and does better in the later summer and so I sprayed it with a warmth spray, upon which it began to grow normally and the slugs no longer attacked it. With the warmth the etheric could once again begin to flow and hence the balance was restored. It is not simply a case of restimulating the etheric, it is more about re-establishing the conditions which will allow the etheric to flow naturally. Both the nature of the pant and the nature of the insect need to be investigated and considered before a successful biodynamic solution using just the Biodynamic preparations can be used.

Mammalian Pests

Mammals such as rodents, possums, rabbits and even deer, can be seen in a similar way however slightly different forces are active. Again I refer you to Dr Kolisko's work on animals to gain a picture of what your particular pest is energetically and therefore what it is doing in your environment.

Being mammals, these animals have fully developed organ systems and are relatively close cousins to Humans. Therefore we must look more to the working of the Ego or World spirit in their activity rather than just the Etheric / Astral balance. The Ego Astral balance may also need to be considered.

If these pests are inside the house, and this goes for ants and cockroaches as well, we must look to the inhabitants of the house for the cause of the infestation. You will find the infestation most directly effects the person with the imbalance. Eg rats in the ceiling above a persons office or bedroom. In these cases while the spray can be applied around and inside the house it can also be given to good effect to the person who has the corresponding imbalance.

Fungi

Dr Steiner gave clear indications regarding fungal attacks, and commented that we can expect to be able to use a general remedy to effect many types of fungus on many types of plant with the same remedies. He suggested that 'rotting' fungal attack was an overly active etheric activity, which sees an activity that is at home in the soil, being drawn up into the plant sphere, due to a satisfactory environment existing in that region. He suggested we needed to draw the etheric back into the Earth and plant and that this could be achieved using 505 (Oak Bark). We can also push the etheric back into the Earth with Silica in the form of Equisetum or horsetail. Clay is also useful in propelling the Cosmic Forces upward to push through a too strong Earthly Substance activity.

However with Powdery Mildew, which is a renowned dry hot fungus, it does not easily fit into this description. In the medical lectures RS suggests Cinnabar — Mercury Sulphate as a remedy. (see my '5 Questions in BD Ag')

Part of the solution for all fungus is to change the physical circumstances as much as possible, which allow for a rampant etheric to exist in the first place. This can be caused by over fertilisation , especially with nitrogen. It can be caused by low light and low air movement or a wet season.

I have looked into the seasonal cycle influences more in recent years and have come to looking at the movement of the Earth's bodies through the seasons and how their connection or not to the internal activities of the plant at the point of these changes manifest as disease manifestations.

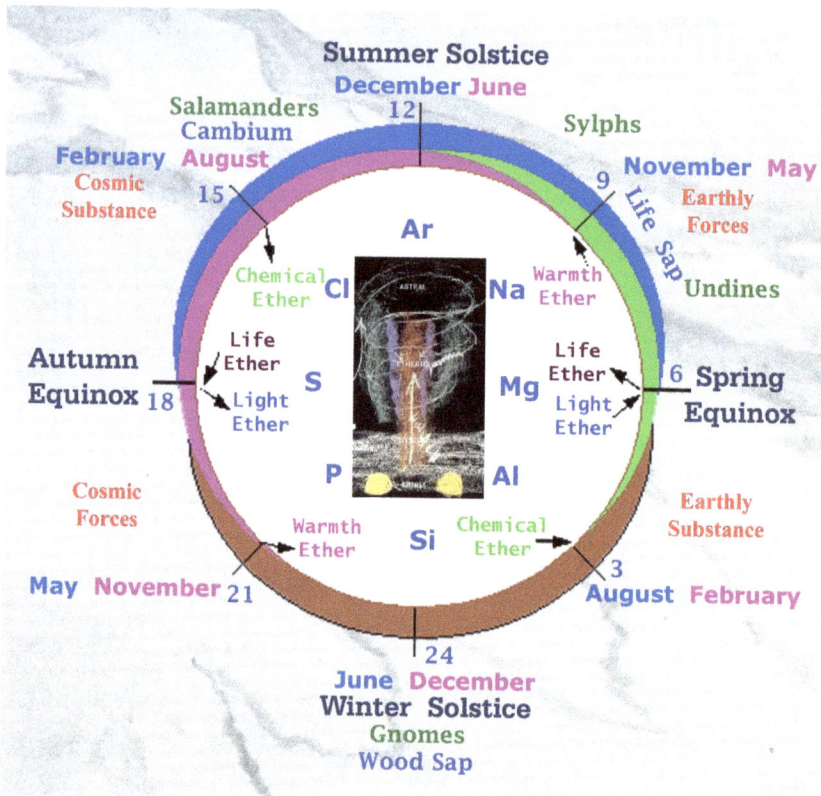

Summer Solstice
December June
Salamanders 12
Cambium
Sylphs

February August
Cosmic 15
Substance
November May
Earthly
Forces

Life Sap

Ar

Chemical Cl
Ether

Na Warmth
Ether

Undines

Life
Ether
S
Light
Ether

Autumn
Equinox 18

Life
Ether
Mg
Light
Ether

6 Spring
Equinox

P

Al

Cosmic
Forces

Earthly
Substance

May November 21

Warmth
Ether
Si
Chemical
Ether

3
August February

24
June December
Winter Solstice
Gnomes
Wood Sap

Bacteria and Virus

If Fungi are an imbalanced Etheric, then bacteria could be an imbalance with the astrality while Virus are an imbalance with the World spirit activity. It might be more appropriate to talk of these as too active or uncontrolled light and warmth ethers, within the etheric sphere of the plant. More recent work using Walter Russels sub atomic elements suggest remedies may exist there.

Peppering

This is a method where the ash of a seed, insect or skin of a mammal is burnt and then spread about the area to be protected. I use it with success for my commercial clients. Its main drawback is that it does not 'close the hole' that the pest came into, and so another like species will come and fill the hole till it is closed.

Peppering appears to be district specific. So possum remedy from one district may not work on an area 100kms away.

A SUGGESTION FOR PLANT OBSERVATION

How can we identify the basic energies of each plant?

When observing plants look for the different layers of influence.

Firstly the species indicates a planetary association. Is it a conifer (Saturn) fruit tree (Jupiter) or shrub (Mars) for instance. - The Ego level

The subspecies indicates the changes within the species .

Is the form of the conifer type - tight (Saturn) , open (Venus), branches facing upwards (outer planet) or down (inner planet)? Is there harmony in the form? (Venus) Are the leaves needles sharp? (Saturn, Mars)or soft (Moon)? almost leaves etc. - The Astral level

What Ethers are working? What is the shape of the leaves? What is the quality of light and spacing of the leaves? Are there flowers? How do they sit in relation to the leaves? Above the plant, or in beneath the canopy? Does any one part of the plant dominate. - The Etheric level

Then there are the localised influences on planting position, soil type and relationship to other trees. - The Physical level.

THE BIODYNAMIC GYROSCOPE

Dr Steiner's worldview, is difficult to develop, comprehend and work with, due to the mutli - dimensional nature, of the holographic spherical reality, we have in front of us. I joke it takes the first 30 years to gain this overview before one can begin to really work with it. Not only are we given four energetic activities to consider, but we also have to consider how these individual energetic activities act when they are occurring within any other sphere. For example we have the astral activities being an externalised activity in the solar system, yet this activity internalises into bodies, and comes into interaction with the physical and etheric and spirit bodily activities. The astrality for instance, has a particular interest in the working of the kidney bladder system and has taken hold of various etheric and physical activities to achieve the job it wants done, with this organ. So we are confronted with the way the astrality is working within an etheric realm, just as much as we have to consider the etheric working in an astral sphere. These complexes of relationships, occur both on the external and internal spheres of activity.

To aid in the comprehension of this enormous picture, I have taken to drawing diagrams, to illustrate the parts. The accompanying diagram is the result of placing all the parts Dr Steiner talked about in his 'Agriculture" Course, onto the

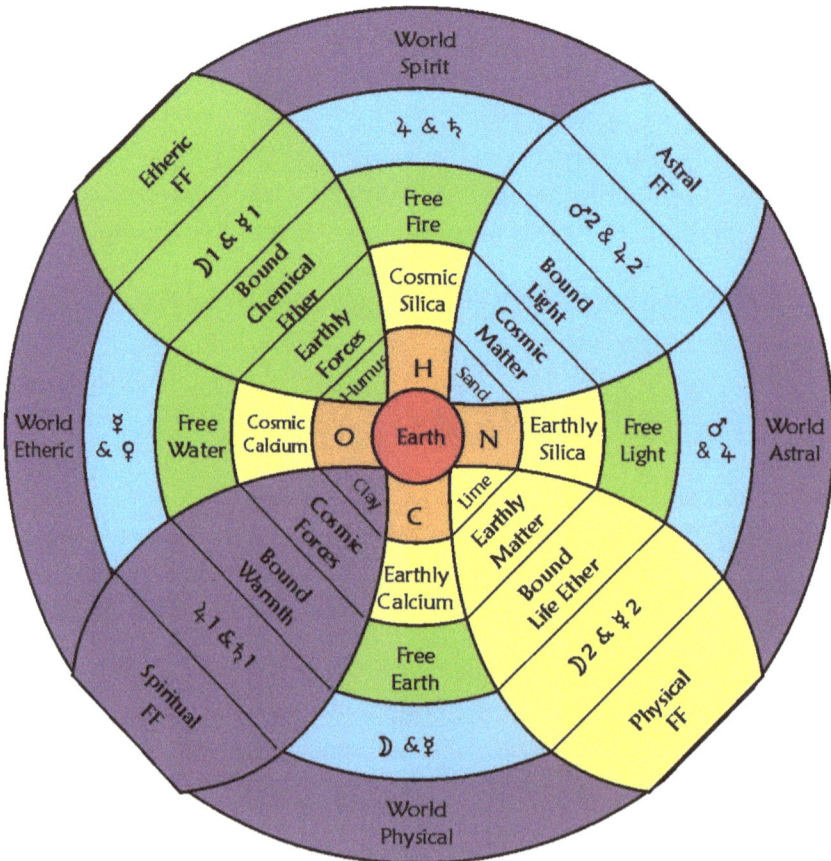

R. Steiner's Agriculture Course
Glen Atkinson

external and internal cross axis of a circle.

The different colours on the external axis should also be on the corresponding circles of the internalised arms, however this makes for more visual confusion, so I have chosen to do away with this detail. The Green colour, seen on the four external arms, representing the layer of the four fold Etheric body, could be joined up, thus showing the influence of the different energetic activities within the internalised bodies as well. So the Green area of the internalised axis, is the zone /arm of the internalised Etheric body, however its various segments indicate how the other energetic activities play into this energetic region.

One of the problems, to overcome in the development of this diagram, was how to allocate the internal arms. It would appear reasonable that the internal arms would be directly related to the external arms, it was just a case of which way to spin the arms for the 'right' relationships. There were many considerations that influenced my final outcome. In the end all the evidence pointed to spin the axis, one step anti clockwise. This makes the internalised bodies - Ego / Spirit and Astral working from above the horizontal axis and the Etheric and Physical bodies from below the horizontal axis. This at first seems to be a somewhat random choice, however as this 'story' continues, the relevance of this placement provides some proof of it correctness.

In this representation the appropriate references from the Agriculture course are placed. Each of these are the carriers of a particular energetic activity in a particular zone. E.g. The Green external ring has places for the free atmospheric elements (and their accompanying ethers) for Water Earth, Air and Warmth, while the internalised areas on this same circle is the position of the bound chemical, Life, Light and Warmth ethers.

These zones represent the 'physical' carriers of the energetic activity of that part of the gyroscopic sphere.

The 'Free Earth' is on the etheric ring of the external physical arm. This is the place where the World Etheric works into the World Physical sphere. This could manifest as water running through the soil and the activation of soil   texture conditions which other life processes may exploit.

The Free Water is the Etheric ring of the World Etheric arm, so this is the most purely Etheric aspect of the gyroscope and best expressed as the Water on the planet, and the life processes this facilitate.   The Bound Chemical Ether, is in the area of the World Etheric working upon the internalised Etheric. This would be a position that strongly supports the life and growth processes in any living being. It would be a significant point for the basis of very good health.

And so on. Each position on the gyroscope can be identified in this manner.

I appreciate, that the path we have taken so far could be considered something of an artistic, rather than scientific process. We have looked at what is the basis of creation, the gyroscopic form, and as Dr Steiner suggested, looked at his description of life according to its big rules. In itself the resulting diagrams are merely a representation, that does not need to have any relevance beyond allowing an objective form for Dr Steiner's concepts. This is what I set out to achieve.

As time went on and in particular, as I studied chemistry. I came to see that this diagram could be much more than a simple work of art, since it is built upon a cosmic form, the form upon which our life intimately depends.

THE GYROSCOPIC PERIODIC TABLE
Circa 1998

The Gyroscopic Periodic Table has been a natural progression of the Astrological and Biodynamic models presented earlier in this work. It is developed from my study of Soil Science, Dr Steiner's Agriculture and Medical Courses, Astronomy and Astrology. I do not pretend to have an extensive knowledge of chemistry and offer the accompanying diagram, as a beginning of what I hope could become an ongoing discussion of both the questions and answers this diagram presents. A more in depth study of this is presented in Glenological Chemistry book.

The process of coming to this diagram, is primarily artistic, and as a practical approach to objectifying Dr Steiner's ideas into a pictorial form, to easily comprehend the holistic approach he suggests. I found that Dr Steiner's world view and its expression, uses the same basis as Astrology; so when I investigated the Periodic Table, and found its basic structure is the same as the Gyroscopic Biodynamic world view, it seemed a natural next step to join them all together. This supplies a very solid base from which to primarily observe and ask questions. I have not yet come across a work suggesting the energetic activity of all the elements of the Periodic Table. So the information this diagram suggests must firstly be taken as indicative. Yet based upon Realities. This is **a challenging question generator from the Universe.** It provides a suggestion for **the energetic activity of every chemical element**. My explorations to date, have convinced me of it usefulness, enough to believe it is worth making available to others, who might like to explore the chemical realm as influences of the energetic body interaction.

It is from this spirit of exploration and the desire for the worth of this diagram to be deepened or revealed as worthless, that I make it freely available to all who would like to take part in its unveiling.

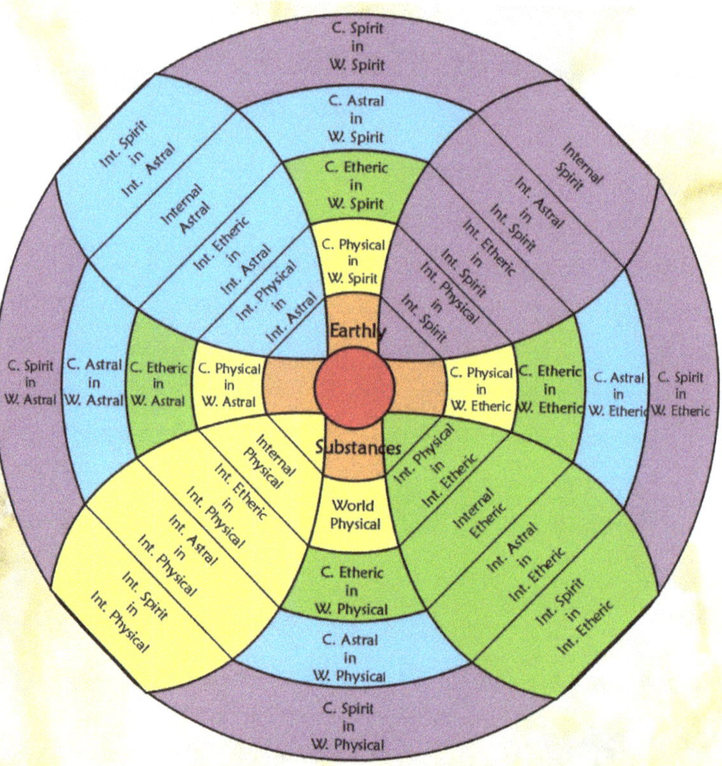

Top diagram (outer to inner sectors):

C. Spirit in W. Spirit
C. Astral in W. Spirit
C. Etheric in W. Spirit
C. Physical in W. Spirit
Int. Spirit in Int. Astral
Internal Astral
Int. Etheric in Int. Astral
Int. Physical in Int. Astral
Internal Spirit
Int. Astral in Int. Spirit
Int. Etheric in Int. Spirit
Int. Physical in Int. Spirit

C. Spirit in W. Astral
C. Astral in W. Astral
C. Etheric in W. Astral
C. Physical in W. Astral
Earthly
C. Physical in W. Etheric
C. Etheric in W. Etheric
C. Astral in W. Etheric
C. Spirit in W. Etheric

Substances

Int. Physical
Internal Physical
Int. Etheric in Int. Physical
Int. Astral in Int. Physical
Int. Spirit in Int. Physical
World Physical
Int. Physical in Int. Etheric
Internal Etheric
Int. Astral in Int. Etheric
Int. Spirit in Int. Etheric

C. Etheric in W. Physical
C. Astral in W. Physical
C. Spirit in W. Physical

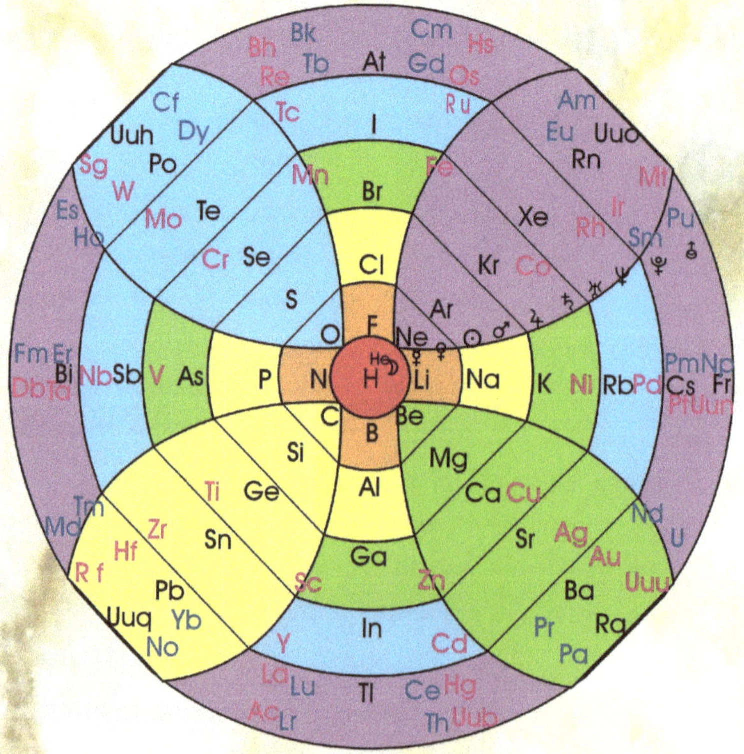

Bottom diagram (element symbols):

Bk, Cm, Hs, Bh, Tb, At, Gd, Os, Re, Ru, Tc, I, Cf, Dy, Am, Uuh, Eu, Uuo, Sg, Po, Rn, W, Mn, Fe, Mt, Es, Mo, Te, Xe, Ir, Pu, Ho, Cr, Se, Kr, Co, Rh, Sm, S, Br, Cl, Ar, O, F, Ne, He, Fm, Er, Bi, Nb, Sb, V, As, P, N, H, Li, Na, K, Ni, Rb, Pd, Cs, Fr, Db, Ta, Pm, Np, Pu, Uup, C, Be, B, Si, Mg, Ti, Ge, Al, Ca, Cu, Nd, Tm, Zr, Sn, Ga, Zn, Sr, Ag, Au, U, Mo, Hf, Sc, Ba, Uuu, Rf, Pb, Yb, In, Cd, Pr, Ra, Uuq, No, Y, Tl, Ce, Hg, La, Lu, Ac, Lr, Th, Uub, Pa

Glenological Chemistry
North

The key question with the Periodic Table is " **What are we being asked to imagine when looking at the rectangular Periodic Table?"**

We are being asked to imagine atoms with a nucleus and with up to seven rings of electrons. In this 'circular picture' we are told there are eight primary arms of elements. This eight arm structure provides the basis for the gyroscopic model we have met earlier from Astronomy and the 7 rings are the same as the six rings I have presented here. The '7th ring' arises as a two fold division of the 6th Zodiacal ring of my diagram. In the Biodynamic Vortex picture this 6th ring has two parts indicated.

All previous associations made in this book can be applied to the Periodic Table and we have **the energetic activity of every chemical element.**

Since first producing this chapter, I have provided a more extensive exploration of the Gyroscopic Periodic Table, as a book "Glenological Chemistry". It is available in hardcopy or for free from www.garudabd.org website, In 'books' .

Biodynamic Plant Growth Achievements
The Significant Moments

1982   Began working with potentised 500 and 501
Broccoli came to head , was then forced back to leaf, then to head again. Providing four subsidiary heads of 5 inch across for each. The center stalk began to rot and smell badly so the trial stopped
It was also noticed white butterfly damage was minimal
A tamarillo was planted under the eve of the house to avoid the frost, however it would not ripen in the season available. Several sprays of potentised 501 were applied. Abnormal leaf differentiation was experienced on three different occasions and the tamarillos were ripened before the frost in subsequent seasons
Began taking potentised 500 & 501 personally.

1983   A half acre of potatoes were grown using only potentised 500 & 501. No Phytophera Infestans was experienced.

1988   Began working with the other BD compost preps in a potentised form.
Lettuce were grown through the winter on wet clay without slug damage.
Brassica were grown through the summer without slug or white but terfly damage.

1989   A project was established to have a hanging basket of 4 'Dizzy Lizzies" always flowering.
This was achieved. As part of the project I would move the plants f rom being in leaf and no flowers to full flower and back again at will and individually. Showing where the spray landed is where it effected
First time a human was treated to move a mouse out of their house.

1990   Bird spray was applied to part of a ripening Loquat tree. The sprayed fruit ripened quicker than the rest, yet the birds left them alone and ate the other loquat on the rest of the tree.
Crickets were moved away from the house. Any in the sprayed area did not make a noise.
Rats moved away from the house and replaced by native rats who lived in the garden.
Chased ants around a caravan finally moving them out with broad band spraying.

1991   Flowering Spray on beans produced three times the flowers and fruit than the control.

1992   Apple trials showing control of Black Spot and Codlin Moth. But not simultaneously.
Worked with Nqatarawa Wines on Flowering, and Ripening. Over six varieties the worst we achieved was a 15% increase in yield.
Phylloxera was reversed, so the new vines effected were indistinguishable from the healthy ones. Our treatments

stopped and the vines reverted.

Neutralised the smell of four dogs in kennels next to a packing shed with one spray of BD preps a week.

Cattle farmer talked of clover leaves the size of bottle tops after spraying Etherics 1000

1993   Sprayed the EDRI factory and surrounding land for possums and rats using BD preps through a motorised backpack sprayer. The resident rats left the next day as did the possums and neither were seen around the area for three months.

During a warm June spell a rose began to mildly shoot. Warmth and flowering sprays were applied and within a month a flower bud was ready to open in July. Only the part of the plants sprayed responded. It took three or so months for the rest of the plant to catch. The following summer those plants flowered the most they had done in the locals memory.

1996   Sprayed every second paddock of a BOP dairy farm with Etherics 1000 at varying rate. The grass growth in the strip were significantly different to each other. Clover density and size were very indicative.

501 comp was also applied to the paddocks. After each grazing in one of the sprayed paddocks the milk vat levels increased by 10%. There was significant amounts of grass left after each grazing in the sprayed paddocks while the unsprayed paddocks were eaten out by mid way through the grazing period. This farm went from the worst farm in the district in June to being one of the best by November. Grant Patton worked on the soil science and nutrition and the credit for this is his.

1997   Made a compost garden at Pukehina. Bought lettuces in a punnets. Sprayed them with 'Pest protection' for slugs. 95% of the seedlings survived and were eaten slug free. I got 40 organic seedlings from a friend and planted them in the garden with no Pest Protection and they were all eaten by slugs. Re planted with punnets sprayed with PP and all were protected.

1998   Kale flowering trial. Three kale all from the same punnet planted next to each other in a triangle. One is sprayed with flowering spray. Two with Leafing spray. Flowering plant flowers, other done. After a time one of the leaf plants are sprayed with flowering and it flowers, other one does not flower for two seasons. A fourth kale plant did not flower for three seasons before dying.

1 ha of Echinancea Purpurea was grown in very low lying land near Edgecumbe. Fungal diseases experienced in nearby paddocks did not occur. Weights and potency were adequate.

1999   Rats moved away from the house for periods of three months before a resident cat arrived.

2000   Rabbits off beans, and tomatoes flowering with the same spray in Pittsburgh USA . Pictures at the website

Capsicums bought to fruiting and full size with no rots, in March after nothing all season. This was done twice.

Pumpkins kept stored on top of each other in the weather under bush till October the following year. The neighbours pumpkins, when stored perfectly, rotted in two months. This was a bad growing season.
Passionfruit flowering controlled.
Commercial glasshouse beans grown with flowering spray routinely

2001   Echinancea and essences product used on cows for mastitis and cell count levels reduced 100 thousand in a week
Pumpkin flowering and fruit quality influenced

2002   At Okawa , a weak spray of possum preps and the population increased threefold. A heavier dose has moved away all resident possums. Some still passing through. Effect lasted up to six weeks after a spray application

HortResearch testing showed proven efficacy of ThermoMax, BirdScare and PhotoMax products.

2003   ZeroIn shown to reduce water uptake in grapes, reducing splitting. Similar results in plums

2004   Cats removed from sprayed area for around three weeks per application.

2005   ZeroIn sprayed on celery to stop upward bolting seeding process.

2007   In UK, Etherics 7 in sludge tanks— reduced crusting and smell, when spread " I usually put five rows into one before the forage harvester goes along. In the E7 field I can only get four rows into one because the crop is heavier"

In the Hague a 3 year trial for Horse Chestnut disease, with BactMin and Etherics 7. Showed effect on the disease — bark damage healed, buds larger. Sprayed areas of the trees held their leaves for 6 weeks longer in the autumn.
BactMin on potaotes showed larger numbers of large potatoes in the treated plants

2010   In Chile Etherics 7 used for drought on turnips. One application at 3rd leaf stage and the harvest was double the size of the unsprayed and less insect damage

2010   RabbitChase on Potatoes worked well

2011   Saltmin trials on grapes reduced salt in new tip growth by 90%

2020   FG4 helps with fungal control, varicose on Lemons, brown rot on peaches, blackspot on apples, powdery mildew on grapes and apples, botrytis on grapes

2023   Sodium Sulphate controlled curly leaf on peaches.

See https://garudabd.org/case-studies/ for 'the evidence' for these stories.

KIWIFRUIT

1993   Began working with Kiwifruit orchards in Bay of Plenty
Etherics 1000 being applied around 4 times per season. Good
changeover experiences .

1996   Began using 'peppers' in the kiwifruit orchards with good success
Some 70 commercial orchards are sprayed each season with these
peppers for Passion Vine Hopper, Scale insect, Fullers Rose Weevil.

1999   Began flowering spray applications to BOP orchard with good
results. Flowers counted conclude this orchard had the most
flowers in their area and that it was the only orchard to have   more
flowers in that season, than the one before.

2002   Ripening spray showed a 1 pt brix raise when sprayed in the last
week before harvest.
ThermoMax success for frost, which continues.

2003   Flowering trials showed a 55% bud break as opposed to 35% on
unsprayed areas.

2005   ZeroIn shows a .7% increase in Brix and 30% increase in TZI, on
Gold Kiwifruit, after 10 days rain,

2007   PhotoMax showed increases in DM (2 pts), size and payout (TZG
double)

CRICKLEWOOD - Graham Reid

1991 -> Began working with Cricklewood farm in Te Puke
Some 20 acres of brassica were grown through a hot summer in poor
sand soil with no white butterfly damage. (Pictures on website).
The first rooting compound was made to strengthen root growth.
Watermelon were sprayed and produced fully ripe 'large apple' sized
watermelon (4" across) , which were sold as novelties   for $1   ea.
(captured on video)
Aphids moved off celery in 12 hours
Anthracnose controlled on melons.
Rooting compound was routinely applied to nursery crops which
lead to quicker growing trees when planted out. (Buyer
comments)
Rabbit and Bird sprays made for sweet corn crops.
White fly were removed from mature Tamarillos trees and stayed
away for four seasons till the trees were removed. When new
plantings were made in 2002 white flys were again removed.
Glasshouse table grapes produced annually with very few fungal or
pest problems.
ThermoMax has been used for 20+ years for Frost. Best achievement
was protected against – 6C. Typically –3C frosts protected.

HUMAN

There has been many experiences of Humans being positively
influenced by the BD preps from simple colds , headaches, emotional
traumas, Hayfever, Asthma, the negative effects of hallucinogenic
drugs, many digestive and metabolic disorders.

And still looking.

BIBLIOGRAPHY

1) Agriculture   Rudolf Steiner
2) The Cycles of Heaven   G. L. Playfair , S. Hill
3) Astrology , Biodynamics & Greenhouse Gases   Glen Atkinson
4) Occult Science - An outline   Rudolf Steiner
5) The Four Ethers   E. Marti
6) Energetic Science and Medicine   Rudolf Steiner
7) Earth & Man   Karl Konig
8) The working of the planets & the life process in Man and Earth.
     B.Lievegeod
9) Man on the Threshold   B. Lievegoed
10) Healing Plants   Wilhelm Pelikan
11) Working on the land with the Constellations   M. Thun
12) Turning up the Heat   Fred Pearce
13) Living Water   Olaf Alexandersson
14) Cycles of Becoming   Alexander Ruperti
15) Sensitive Chaos   T. Schweck
16) Insect and Plant   K. Konig
17) Moon & Plant   Agnes Fyfe
18) Fundamentals of Anthroposophical Medicine   R. Steiner
19) Energetic Relations in the Human Organism.   R Steiner
20) Psychology and the Four Elements   S. Arroyo
21) The Twelve Groups of Animals   E. Kolisko
22) Hermetic Astrology   R. Powell
23) Astronomical Letters   E. Vreede
24) The Nature of Substance   R Hauschka
25) Agriculture of Tomorrow   E & L Kolisko

Biodynamics Decoded

CAPRICORN
Cardinal, Earth, Saturn 2
Limb Metabolic
Root Plants
eg Seeds of Carrots
Arthopoda - Insects
Aluminium

PISCES
Mutable, Water, Jupiter 1
Rhythmic
Leafing Plant
eg. Leaves of Cabbage
Mollusca - Shellfish
Chlorine

SCORPIO
Fixed, Water Mars 2
Nerve Sense
Leafing Plants
eg Roots of Cabbage
Amphibia - Frogs
Carbon

TAURUS
Fixed, Earth, Venus 1
Nerve Sense
Rooting Plants
eg Roots of Carrots
Echinodermata - Starfish
Nitrogen

VIRGO
Mutable, Earth, Mercury 2
Rhythmic
Rooting plants
eg Leaves of Carrots
Aves - Birds
Sodium

CANCER
Cardinal, Water, Moon 1
Limb Metabolic
Leaf plants
eg Fruit of Cabbages
Protozoa - Bacteria
Phosphorus

AQUARIUS
Fixed, Air, Saturn 1
Nerve Sense,
Flowering Plants
eg The roots of Roses
Vermes - Worms
Oxygen

SAGITTARIUS
Mutable, Fire, Jupiter 2
Rhythmic
Fruiting Plants
eg Leafs of the lemon tree
Pisces - Fishes
Magnesia

ARIES
Cardinal, Fire, Mars 1
Limb Metabolic
Fruiting Plants
eg Fruit of Peaches
Tunicata –Sea Squirts
Silica

LIBRA
Cardinal, Air, Venus 2
Limb Metabolic
Flower plants
eg Flowers of Roses
Reptilia - Lizards
Calcium

GEMINI
Mutable, Air, Mercury 1
Rhythmic
Flowering plants
eg Leaves of Roses
Coelentrata - Corals
Sulphur

LEO
Fixed, Fire, Sun 2
Nerve Sense,
Fruiting plants
eg Roots of Lemon trees
Mammals - Cows
Hydrogen

SATURN
Prep 507
Valerian tincture
Element - Phosphorus
Organ - Spleen
Metal - Lead
Strengthens Spirit

JUPITER
Prep 506
Dandelion/Mesentry
Element - Hydrogen
Organ - Liver
Metal - Tin
Helps Spirit and
Physical entwine

MARS
Prep 504
Nettle / Earth
Element - Nitrogen
Organ - Gall bladder
Metal - Iron
Harmonises the
Astral body

VENUS
Prep 502
Yarrow / Stag bladder
Element - Sulphur
Organ - Kidneys
Metal - Copper
Opens the Etheric
to the Astral

MERCURY
Prep 503
Chamomile /Intestines
Element - Oxygen
Organ - Intestines,
Metal-Mercury
Strengthens the
Etheric body

MOON
Prep 505
Oak Bark / Skull
Element - Carbon
Organ - Reproductive
Metal - Silver
Retards a
rampant Etheric

FIRE
Active Intuitive
Spirit - Will - Warmth ether
Seed & Fruit - Spheres
Adult, Hydrogen, Clay
Aries, Leo, Sagittarius
Cosmic Silica, Cosmic Forces

AIR
Rational Intellectual
Astral Body - Light ether
Flower - Pointed leaves
Pupa, Nitrogen, Sand
Gemini, Libra, Aquarius
Ter. Silica, Cosmic Matter

WATER
Nurturing, Emotional
Etheric Body - Chemical
Leaf & stems -Hemisphere
Grub, Oxygen, Humus
Cancer, Scorpio, Pisces
Cosmic Calcium, Earthly Forces

EARTH
Sustaining, Practical
Physical Body - Life ether
Root - Squares
Egg, Carbon, Lime
Taurus, Virgo, Capricorn
Ter. Calcium, Earthly Matter

FIXED
Consolidating - Antithesis
Nerve Sense system
The Head
Soil and Root
Salt - Basalt

MUTABLE
Changeable - Synthesis
Rhythmic system
Abdomen - Lungs & Heart
Leaf region
Mercury - Bentonite

CARDINAL
Initiating, Leading - Thesis
Limb Metabolic system
Digestion & Limbs
Flower & Fruit
Sulph - Sulphur

SUN +
Prep 501 - Silica / Horn
Spring & Summer
Dawn till Midday
Catabolic, Nutritive, Cold
Externally expansive and warm
Strengthens the upright growth
of the plant & enhances
the Astral & Spirit
Sun, Outer planets, Silica

MOON -
Prep 500 - Cow manure / Horn
Autumn & Winter
Afternoon & Evening
Anabolic, Reproductive, Warm
Externally contractive and cold
Enlivens the soil and growth
through the enhancing of the
Etheric and Physical bodies
Moon, Inner planets, Calcium

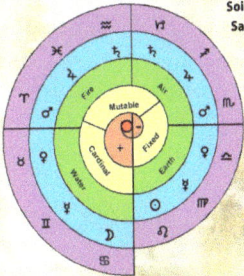

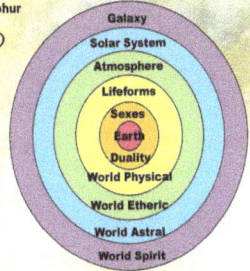

Galaxy
Solar System
Atmosphere
Lifeforms
Sexes
Earth
Duality
World Physical
World Etheric
World Astral
World Spirit

UNITY
Primal Oneness – Earth
This level is the manifest Earth
and the spiritual state of unity
from which all life begins and
returns. The atom acts
as the doorway between
spirit and matter

The references on this diagram
are drawn from the work of
Dr R Steiner, Dr R Hauschka
Dr E Kolisko , Dr Lievegeod
& Glen Atkinson

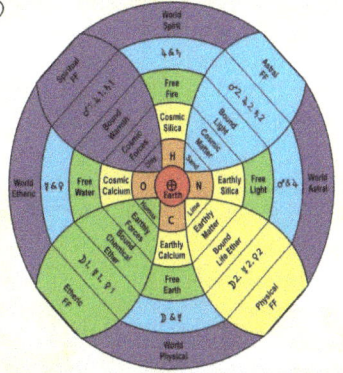

www.garudabd.org
© Garuda Trust 2006